Green Energy and Technology

Climate change, environmental impact and the limited natural resources urge scientific research and novel technical solutions. The monograph series Green Energy and Technology serves as a publishing platform for scientific and technological approaches to "green"—i.e. environmentally friendly and sustainable—technologies. While a focus lies on energy and power supply, it also covers "green" solutions in industrial engineering and engineering design. Green Energy and Technology addresses researchers, advanced students, technical consultants as well as decision makers in industries and politics. Hence, the level of presentation spans from instructional to highly technical.

Indexed in Scopus.

Indexed in Ei Compendex.

Long Shi · Haihua Zhang

Solar Chimney Applications in Buildings

For Engineers and Scientists

Springer

Long Shi
State Key Laboratory of Fire Science
University of Science and Technology
of China
Hefei, Anhui, China

Haihua Zhang
School of Engineering
RMIT University
Melbourne, VIC, Australia

ISSN 1865-3529 ISSN 1865-3537 (electronic)
Green Energy and Technology
ISBN 978-3-031-45217-8 ISBN 978-3-031-45218-5 (eBook)
https://doi.org/10.1007/978-3-031-45218-5

This Springer imprint is published by the registered company Springer Nature Switzerland AG
The registered company address is: Gewerbestrasse 11, 6330 Cham, Switzerland

Acknowledgements

This research was supported by the Australian Government through the Australian Research Council (DE200100892), the National Natural Science Foundation of China (No. 52274235), the Fundamental Research Funds for the Central Universities (No. WK2320000056), and the Natural Science Foundation of Sichuan Province (No. 2022NSFSC0278). We also would like to acknowledge the contribution from Professor Guomin (Kevin) Zhang, Dr. Yao Tao, Professor Vivian W.Y. Tam, Professor Jie Li, Professor Sujeeva Setunge, Professor Chuangang Fan, Professor Dong Yang, Associate Professor Kate Nguyen, Associate Professor Fengling Han.

Summary

This book describes in depth the fundamental effects of buoyancy, a key force in driving air, and transporting heat and pollutants around the interior of a building. It is essential reading for anyone involved in designing and operating modern sustainable, energy-efficient buildings, whether a student, researcher, or practitioner. The book presents new principles in natural ventilation design and addresses surprising, little-known natural ventilation phenomena seldom taught in architecture or engineering schools. Despite its scientific and applied mathematics subject, the book is written in simple language. It contains no demanding mathematics covering qualitative and quantitative aspects of ventilation flow analysis. It is, therefore, suitable for non-expert readers who want to develop intuition of natural ventilation design and control (such as architects and students) and those with more expertise whose work involves quantifying flows (such as engineers and building scientists).

Contents

Chapter 1
Introduction and Background

1.1 Historical Overview of Passive Ventilation

As a result of the severe repercussions of global warming, an increasing number of stakeholders have shifted their attention from conventional to renewable energy sources to alleviate the energy load and relieve the strain of high energy prices. Globally, heating, ventilation, and air-conditioning (HVAC) systems have the most significant proportion, 40–60% of the overall amount of energy used by the building sector [1]. With the frequent emergence of extreme global climates and energy crisis, eco-friendly buildings are evolving to face the challenge of climate change. To this environmentally conscious, natural ventilation systems with thermal regulation and indoor air quality improvement have received widespread attention and are proposed as an alternative or assistance to mechanical systems. Currently, most contemporary buildings depend exclusively on mechanical ventilation to maintain an acceptable level of interior comfort. These active systems use a significant amount of energy and take up valuable space owing to their relatively high volume and structural complexity. Mechanically ventilated buildings are often relatively airtight to limit energy use and heat loss, resulting in an insufficient supply of fresh air [2]. Due to passive techniques' economic and energy-saving benefits and their resistance to noise and carbon dioxide emissions, passive systems are increasingly being marketed as low-cost and low-energy choices for energy-efficient buildings [3, 4]. It has been discovered that passive ventilation can offer appropriate capacity for thermal regulation and satisfy indoor air quality in climates where mechanical systems are unnecessary [5–7].

Among the passive design, buoyancy-driven passive solar systems can achieve multiple functions, following different airflow patterns and temperature distributions according to local climate conditions and building features. A thoroughly designed solar-assisted ventilation system exhibits different flow and thermal behaviors, leading to heat dissipation and acquisition, thus having space heating and cooling functions [3, 8]. The solar façade collects and reflects incoming solar energy,

L. Shi and H. Zhang, *Solar Chimney Applications in Buildings*, Green Energy and Technology, https://doi.org/10.1007/978-3-031-45218-5_1

converting direct or indirect solar energy into usable heat. However, the thermal behavior and fluid dynamic properties of solar-assisted stack ventilation only in multiple spaces building under the combined solar radiation and convective heat transfer have not been resolved.

The progress is still slow regarding expanding applications pertaining to multi-storey buildings. With the rapid growth of the global population, the multi-storey building has become an influential form that can release people's living requirements. Previous studies on natural ventilation methods in this area have mainly focused on simulating the energy performance of natural ventilation systems and their combinations numerically or experimentally in a single room, while there are relatively fewer studies on the multiple chambers that employ natural ventilation to minimize energy consumption for multi-storey buildings. Although the numerical simulations can predict the airflow in detail, the computational progress is time-consuming, and the impacts of influential factors on the ventilation capacity are unintuitive. Due to the lack of easy-to-apply mathematical approaches, critical parameters of some building elements cannot be determined through unified specifications, which places barriers to further promoting the use of solar chimneys as a valid ventilation technique for practical multi-storey buildings.

The coupling of SC-assisted natural ventilation for multi-storey buildings remains open. Many existing parameterization models have been discussed extensively in the simple case of a single room or individual air channel. It is difficult to explicitly tackle a general mathematical approach to stack ventilation in multi-zone buildings. The possibility of SC-assisted multi-storey buildings with multiple air inlets and a single outlet complicates the flow process as the dimensionality increases. Contradictory conclusions appear due to differences in study subjects and calculation positions. Therefore, this research aims to develop a related mechanism to predict the ventilation capacity and easily apply it during the design phase for solar chimney-assisted natural ventilation. The interdependency between core variables and ventilation rates is presented by numerical modeling. A mathematical model of a multi-storey building with equal height and connected by an external vertical solar chimney will be developed. The model will be analyzed theoretically and numerically. The airflow characteristics through the building zone and chimney cavity under the uniform solar radiation will be addressed, which can significantly enhance the understanding of stack-driven natural ventilation in multi-storey buildings. This research systematically investigated the impact factors of SC-assisted natural ventilation in multi-chambers on thermal behavior and ventilation rates. The heat transfer behavior and airflow distributions differ from those in an SC alone utilized in single-room buildings.

1.2 The Principles of Natural Ventilation

As passive ventilation strategies are the subject of extensive research, before exploring the potentials of the combined ventilation systems, it is necessary to present and recapitulate an overview of the individual natural ventilation system and the primary principles that cover existing passive approaches. The driven forces control the airflow pattern, temperature, and distribution through a building. According to the sources of pressure difference, natural ventilation means wind-induced, buoyancy-driven, and their combination.

1.2.1 The Basis of Passive Ventilation

Two fundamental natural effects—wind and thermal buoyancy—can be exploited to drive ventilation airflows. The system's inherent characteristics and the uncertainty of natural conditions affect the airflow path, ventilation rate, and temperature change. Air movement, also called natural convection, occurs under pressure differences. Different power resources cause it, including wind (as a momentum-induced airflow), buoyancy effect (produced by internal and external air density differences), or both of them [9], humidity difference, and mechanical power, which exchange stale indoor air and ventilate the interior spaces [10]. Typical natural ventilation systems with purpose-built openings can be classified into four categories [5, 11], including wind-induced ventilation, buoyancy (stack) ventilation, and hybrid (mixed mode) ventilation. Intentional openings include windows, doors, wind catchers, chimneys, and roof vents. Natural ventilation systems are sometimes included in passive cooling methods for dissipation and heat gain operations. However, the systems have many uncertainties, especially under extreme ambient climates. The success of adequate natural ventilation relies on natural forces available, building features, airtightness, and occupant behaviors [12–14]. By a comprehensive review, Oropeza-Pereza et al. [15] interpreted three active and ten passive cooling methods based on availability, reliability, and economic feasibility. The efficiency of different systems was discussed according to energy balance (e.g., internal heat gains and heat transfer). It was reflected that passive methods could reduce indoor temperature. They explored the possible variation of indoor temperature caused by natural ventilation.

1.2.2 Typically Driven Forces

1.2.2.1 Wind-Driven Force

A wind-induced system is a typical way to ventilate space naturally. Under the external wind, the distribution of windward static pressure increases, resulting in a positive or high-pressure region [16, 17]. Vortexes may be produced on the leeward side, which forms a low-pressure or negative region. Under the pressure gradient, the air is driven from the domain with positive pressure to the region with negative pressure. The ventilation intensity is concerned with the area of air intakes/extractors and the magnitude of the wind force, as seen in Table 1.1. The volume control dampers can capture the wind at a high altitude and distribute it through the building and moderate airflow (Fig. 1.1).

1.2.2.2 Buoyancy-Driven Force

The ventilation driven by temperature gradient is called thermal buoyance ventilation. One of the typical airflow paths can refer to Fig. 1.2. The intensity of the chimney effect is associated with the difference in air import and export height and the difference in internal and external air density. When there is any heating source inside the building, the internal air temperature may exceed the external air temperature. The warm air rises to the upper part of the space, leading to an overpressure compared to the outside environment. The air is then driven from indoors to outdoors. The neutral pressure plane (NPL) is essential in calculating the pressure difference. The air pressure between inside and outside air reaches equilibrium at NPL. The NPL could vary from 0.3 to 0.7 of the total building height. An overpressure above NPL exhausts the air through the openings, whereas a negative pressure below NPL drives the air into buildings [21] (see Fig. 1.2).

Besides, the airflow induced by the humidity gradient is like that of the temperature difference. Water vapor is relatively lighter than dry air. Buoyancy induced by humidity is only effective in those locations where the outdoor humidity is very low.

1.2.2.3 Combined Forces

Two natural driving forces mentioned above usually act simultaneously, where sometimes one driving force could be dominant under a particular condition. As Fig. 1.3 illustrates, the combined forces may be mutually exclusive or compensate each other, depending on the incident wind angles and whether the internal or external temperature is high. The airflow rate resulting from the combined effects can be calculated in a root-square function.

Table 1.1 Summary of combined typical natural ventilation systems

Natural ventilation systems		Buoyancy-driven ventilation					Wind-driven ventilation							Hybrid solution		Heat modulation or amortization technique
Architectural elements/other systems		Trombe wall (TW)	Solar (thermal) chimney (SC)	Double-skin façade (DSF)	Atria and courtyard (A/C)	Solar walls with photovoltaic module (PV)	Wind tower (WT)	Windcatcher (WC)	Fenestration (Sigle-sided and cross-ventilation) (F)	Wing walls (WW)	Wind cowls/scoops (WS)	Rotating wind cowl (RC)/turbine ventilators (VT)	Exhaust cowls/roof vents/roof cowls (RV)	Heat recovery ventilation (HRV) devices/heat transfer device (HTD)	Earth-air heat exchanger (EAHE)	Nocturnal cooling applications (NC)
Buoyancy driven ventilation	Trombe wall (TW)		[29, 33, 110]											[89-100, 121]		
	Double-skin façade (DSF)		[41, 122]		[41, 54]	[123]	[27, 124, 125]							[126]		
	Solar (thermal) chimney (SC)				[54, 66]	[127]	[26, 107, 108, 112, 113, 115, 125, 128-130]	[28, 114, 131]	[132, 133]		[134]			[80-88, 103, 135]	[64, 136, 137]	[138, 139]
	Atria and courtyard (A/C)						[27, 116, 140, 141]						[54]			[54]
	Solar walls with PV															
Wind-induced ventilation	Wind tower (WT)											[24]	[142]	[102, 104, 105, 143-148]	[72]	[149-154]
	Windcatcher (WC)									[155]	[155]			[74, 104, 145]		[28]
	Fenestration (Sigle-sided ventilation and cross-ventilation) (F)															
	Wing walls (WW)															
	Wind cowls/scoops (WS)															
	Rotating wind cowl (RC)/turbine ventilators (VT)															
	Exhaust cowls/roof vents/roof cowls (RV)															

Note The shaded part of the table is the duplicated pair-wise comparisons; in order to facilitate the legend annotation of Sect. 1.6, the full name of each type of system is abbreviated in parentheses. Those blank cells represent that no study has been found in the literature regarding the related combination, where future studies are suggested to confirm their viability in practice under various climate conditions

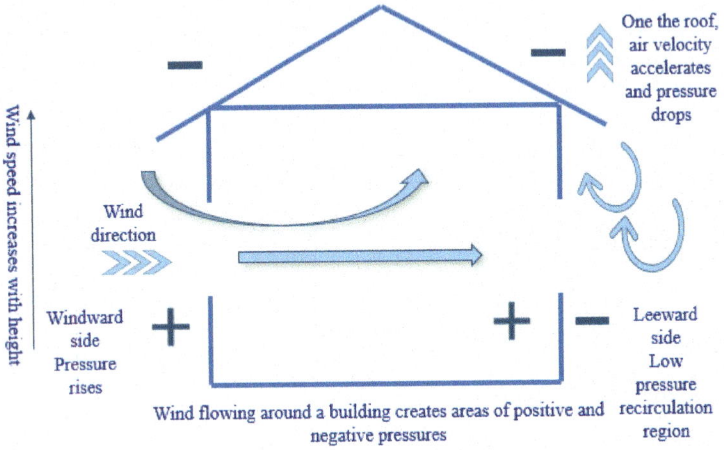

(a)

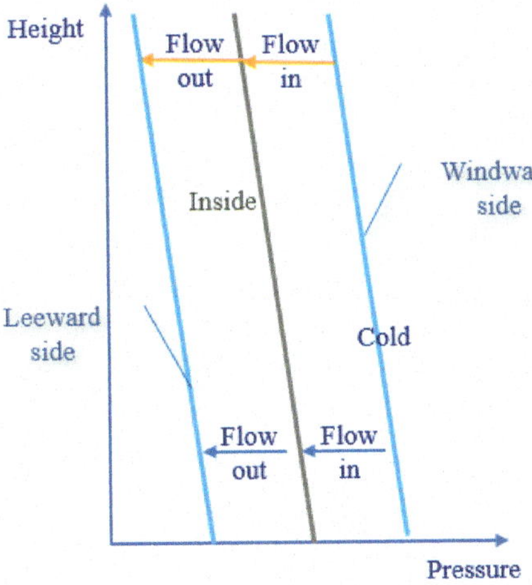

(b)

Fig. 1.1 **a** Wind-driven airflow (pressure differences result in air movement). Revised from [18, 19]; **b** wind action only with the pressure of equal magnitude on windward and leeward sides. Revised from [20]

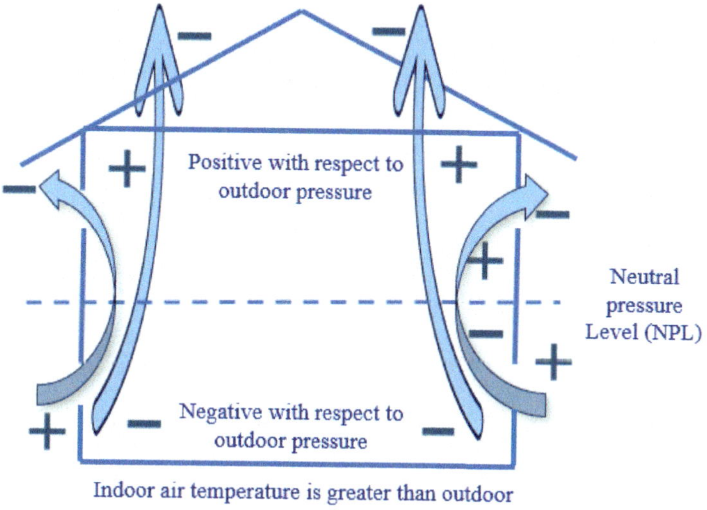

(a)

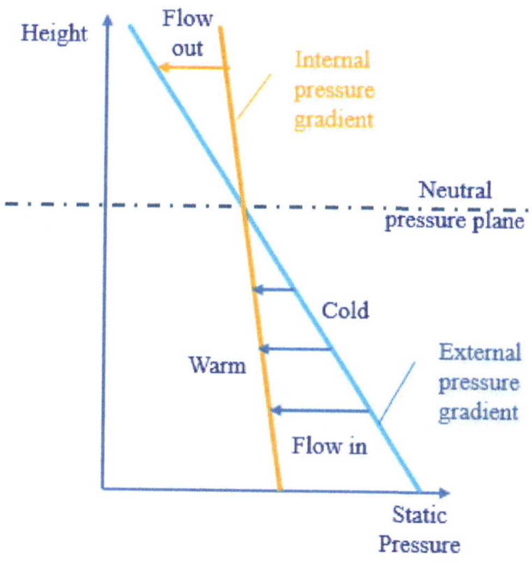

(b)

Fig. 1.2 Airflow caused by the stack effect (density differences result in air mass movement). Revised from [18, 19]; stack effect only with neutral pressure plane [22]

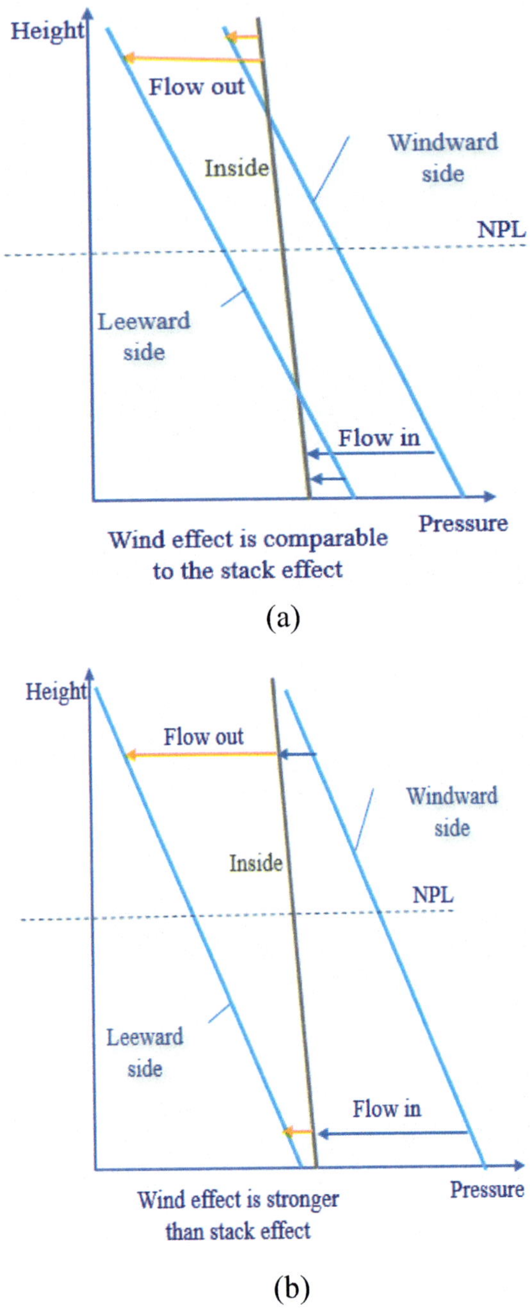

Fig. 1.3 Pressure gradient and airflow patterns of combined effects. Revised from [23]

1.3 Typical Natural Ventilation Approaches

1.3.1 Single-Sided Ventilation and Cross-Ventilation

Openable windows are commonly used vents of natural ventilation systems. As expressed in Fig. 1.4, single-sided ventilation refers to space with one opening connected to the external environment. Air exchange could take place through stack effect and wind pressure.

Naturally ventilated buildings require to be relatively narrow to allow the airflow to penetrate deeply into a building. The adequate depth of a single-open space driven by wind is twice the height from the floor to the ceiling. The adequate depth of the double-opening space driven by buoyancy is 2.5 times the height from the floor to the ceiling. Gan [25] derived a model based on CFD modeling to predict airflow patterns, temperature profiles, and the mean age of air. The model can support an assertion of the efficient depth of air distribution in buildings. The results showed that the maximum depth for adequate air distribution in a single-sided ventilation space is concurrently determined by the opening dimension, internal heat gains, and external air temperature.

Cross-ventilation is with purpose-built openings (two or more openings on the opposite walls) (see Fig. 1.5), using the prevailing wind for indoor ventilation. The

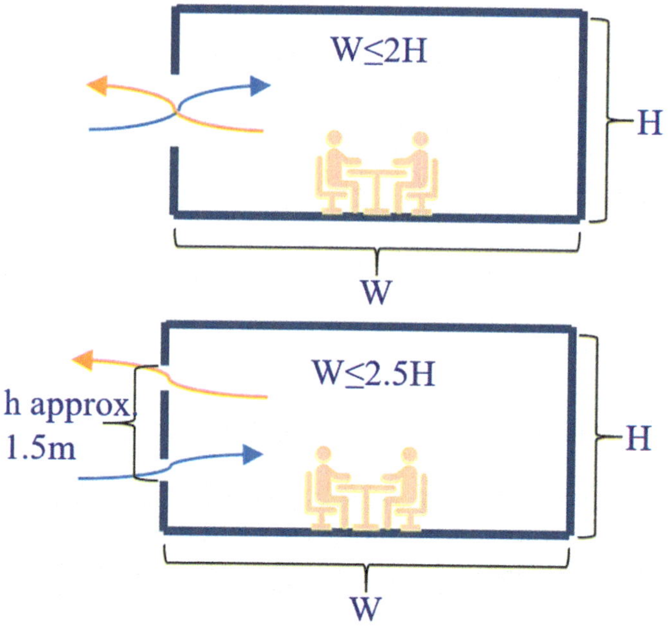

Fig. 1.4 Single-sided ventilation. Revised from [24]

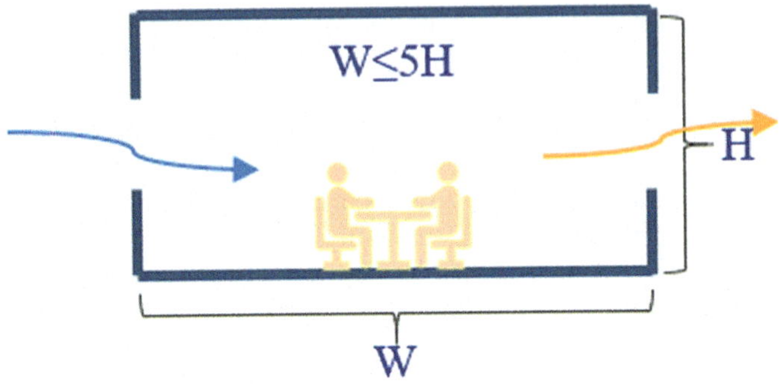

Fig. 1.5 Cross-flow ventilation. Revised from [24]

recommended horizontal depth for cross-ventilation needs to be less than five times the floor-to-ceiling height of the space.

1.3.2 Wind-Induced Ventilation

Wind-driven ventilation can be realized through different systems, such as wind towers/catchers, roof vents, and rotating ventilators. These structures are usually installed on the roof level to enhance the interaction with the external wind. The airflow follows the pressure gradient and exhausts through purpose-built openings or the leeward side (see Fig. 1.6). The size and location of these openings and the distribution of internal walls show a high impact on promoting cross-flow. Khan et al. [26] summarized and classified different wind-driven ventilation techniques and presented their corresponding flow rates. Hughes et al. [27] retrospectively summarized the wind tower systems extensively and evaluated the development and integration potentials of different cooling methods with the commercial windcatcher. Jomehzadeh et al. [28] reviewed previous studies on windcatchers from IAQ and thermal comfort perspectives.

1.3.3 Buoyancy-Driven (Stack) Ventilation

In buoyancy-driven ventilation (see Fig. 1.7), an upward airflow is produced because of the thermal buoyancy under a temperature gradient. One of the attempts to the utilization of stack ventilation methods is to develop solar heating and cooling technologies. Three prerequisites are generally needed to stack ventilation: relatively lower intakes and higher extractors and heat sources (internal or external heat sources). Solar-induced approaches employ direct or indirect solar energy to capture solar irradiance and convert energy into usable heat. Typical solar-driven ventilation

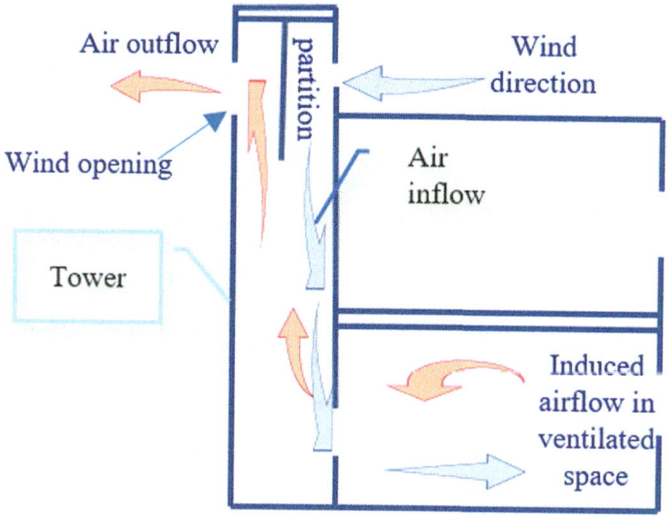

Fig. 1.6 Wind tower

systems which could be attached to the building envelope and mounted on rooftops [29], such as the solar chimney [30], Trombe wall, and double-skin facades (DSF), are in the nature of similar operational concepts using solar-induced thermal convection and buoyancy as they have openable channels that induce airflow circulation under the aid of insolation and preferable in temperate climates [10].

A schematic diagram of a solar chimney presented in Fig. 1.8 consists of a glass wall, an absorber, a tuyere, and a heat-insulating material. Air movement is typically

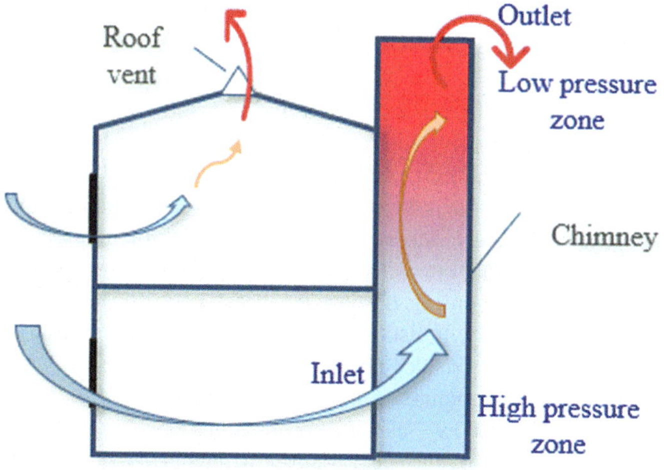

Fig. 1.7 Buoyancy-driven ventilation

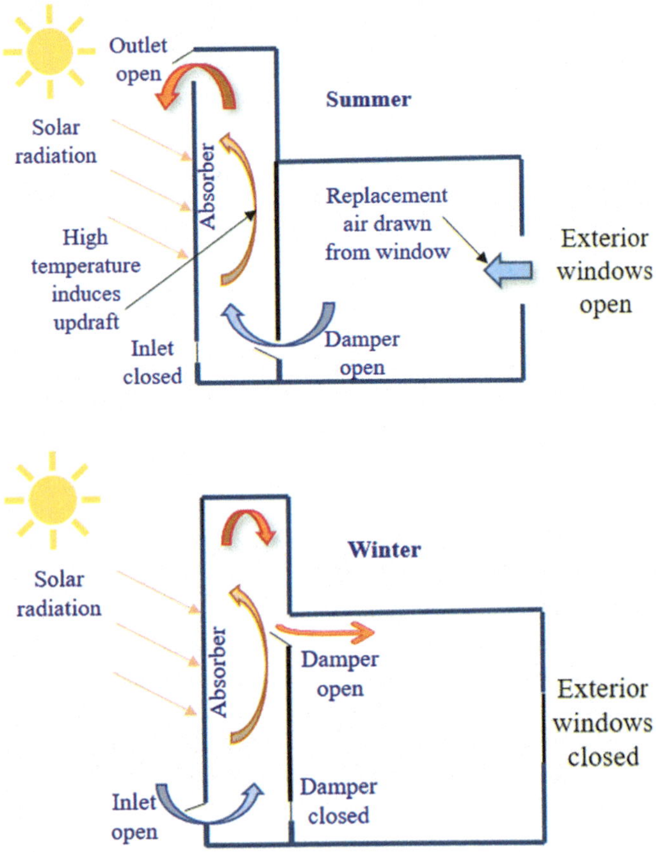

Fig. 1.8 Schematic of solar chimney. Revised from [37]

affected by the intensity difference between indoors and outdoors, resulting from thermal gradient (naturally driven convention) and wind (force convection) [31–34]. Thirugnanasambandam et al. [35] reviewed the thermal and ventilation performance and numerical models of solar-induced methods. It was indicated that air change per hour (ACH) depends on the chimney configurations (e.g., inclination angle, aspect ratio, and inlet size) and solar intensity. Quesada et al. [7, 36] summarized the studies on solar facades explored during the last ten years based on theory, experiment, development, and feasibility. It also reviewed the transparent and translucent solar facades. Not only do solar facades absorb and reflect incident solar radiation, but they also transmit directly obtained solar heat to buildings.

The double-skin façade (DSF) as a heat protection method has become an accessible solution. However, challenges then arise due to the intricacy of thermal stratification, complicated airstream patterns, and the adaptability of DSF to different climatic scenarios [38, 39]. Pérez-Grande et al. [40] explained the airflow mechanisms, the process of solar radiance transmission, and the vertical temperature gradient through the façade layers. The solar incidence that enters through the outer skin is absorbed by the inner skin, which causes an emission of longwave radiation in various directions, resulting in air temperature increases in the air passage [41, 42]. The heat interchange and airflow network involved in the DSF are described in detail in refs. [43, 44]. Excepting solar irradiance, wind characteristics play a pivotal role in the thermal and airflow behavior of the DSF. Wind generates different wind pressures around the DSF. When the wind effect is considered in the study, the wind-driven will be the dominant force in driving air movement through the DSF layers [45, 46].

Atria are typically embedded in buildings, thereby forming centerpieces with glazing space in buildings for natural ventilation purposes [26, 47]. The air circulation through atria spaces is predominantly driven by buoyancy effects [48]. Holford and Hunt [49] advocated raising the temperature of the warm air layer concentrated in the higher part of the space, thereby raising the air pressure of the upper layer. Expanding the depth of the heated air layer is attributed to the enhancement of stack ventilation.

Wind factors, however, can affect the stack ventilation positively or negatively. Horan and Finn [50] described the impact of external wind on the airflow rate inside a double-storey office building under various wind speeds and wind directions through computational fluid dynamics (CFD) study. It is explored that the linear relationship between ACH and wind properties is evident. The energy efficiency of the atrium with the same geometry and glazing materials increases with the height of the building [49, 51–53]. Moosavi et al. [54] explained how building designs differed according to the diversity of building functions and explored the contributing parameters of atriums, intending to obtain prudent designs of the atrium to refine the energy performance of buildings. However, when the excessive heat transfer from the atrium space to adjacent spaces in the hot season and the heat loss caused by the large glass curtain wall and continuous air stratification in the cooling season, the atrium may increase cooling loads to regulate thermal comfort [55, 56]. Also, an unsuitable cavity depth design can cause the above defects [57].

1.3.4 Hybrid Ventilation

Mix-mode ventilation is generally referred to as the coupling of mechanical and multiple passive ventilation in energy-saving buildings. The mechanical systems include forced ventilation (displacement) such as fan-assisted (extraction fans supplement ventilation rate) and heat exchangers [5], which are only intended to use when natural systems are inadequate [58]. A constraint of natural ventilation is the high dependence on ambient climate. When the outdoor wind speed is too meager, the external temperature is too excessive, or the outdoor humidity is extremely high, the accessibility of natural ventilation could then be diminished. Furthermore, a standalone system tolerates the likelihood of excessive undulation in the ventilation rate.

For example, air change caused by the stack effect alone may not generate enough internal air movement in windless conditions.

In contrast, mix-mode ventilation systems extend natural ventilation practices in different building types and diverse climates [59, 60]. They are commonly used to ensure a steady and persistent airflow rate. Yang and Li [61, 62] compared three buoyancy-based ventilation schemes based on a ten-storey building model, including stack ventilation with/without an attached chimney and a stack-based hybrid ventilation system. It developed analytical models and conducted numerical simulations based on CFD, hereafter obtained corresponding ventilation rates and derived energy performance equations. According to the theoretical analysis, dimensionless expressions were derived for proposed systems, and the contributing factors were identified.

1.4 Combined Technologies

1.4.1 Practical Implementations

Table 1.1 summarizes the different combined passive-assisted ventilation systems in a building to enhance ventilation and thermal performances. It was known that relying on a single natural ventilation system may result in under and overventilation relative to a design objective. The combined approaches of passive ventilation systems as an economical solution are then embraced. The two driving forces of a natural ventilation system that produce air circulation vary randomly. Moreover, the difficulties in controlling natural ventilation, with airflow being unpleasant in some places and stagnant in others, result in not working as expected. Therefore, interruption in regular operation may occur during unfavorable climate conditions.

Integrated approaches are advised appreciably economically. When natural ventilation alone cannot meet the required ventilation performance, the scheme of alternatives may be considered. Energy consumption would considerably be reduced by adopting multiple passive strategies [63]. Moreover, the implementation of the practical corporation of passive technologies needs to account for the functions of the buildings.

Since buoyancy-driven natural ventilation systems with vertical air shafts can be easily accommodated into a building structure, they are more accessible to integrate into buildings and work with other natural ventilation facilities to induce high air ventilation rates. The buoyancy provides adequate driving forces to extract the stagnant air from the buildings under intensive solar radiation. By a coordinated integration of components, DSF, solar chimney, and atrium can achieve high-quality feasibility. A few studies [29, 33] have outlined the solar chimneys incorporated with selected systems to enhance ventilation performance. The recommended systems include Trombe walls, earth-to-air heat exchangers, solar roof collectors, evaporative or adsorption cooling components, and the active solar systems of solar chimneys. The research indicated that the optimization and control strategies of such coupling systems deserve further study under various climatic scenarios. Likewise, the atrium can collaborate with the solar chimney or double-skin facade. Implementation of incorporative ventilation techniques may break the limitations of building design. For instance, it is generally recognized that deep surrounding spaces place an obstacle for the buoyancy ventilation, leading to insufficient buoyance effect to drive the air pass through a remote district. Moreover, it is observed that using buoyancy-only ventilation models shows deficient effects on the thermal conditions in extreme climates [29, 33, 54, 64, 65].

Moreover, solar-assisted components integrated into stack ventilation are an efficient way to refine the performance of passive ventilation [33], contributing to the enhancement of the capability of integrated systems. Hussain and Oosthuizen [66] examined the coupling of the atrium and solar chimneys with diverse geometric configurations. It was affirmed [54] that the atrium scheme, coupled with a solar chimney, could be viable to improve the buoyancy effect [40]. It revealed that the application of atrium-independent buildings is efficient in temperate climates with a low impact on thermal stratification [41]. Chan et al. [10] reviewed the previous studies on passive solar technologies for regulating space thermal comfort and indoor environment. The pros and cons of each technique were discussed. It is believed that a combination system that is compatible with space heating and cooling is an acceptable solution for improving natural ventilation.

The role of combined systems in a building can be summarized as follows. First, the combination of multiple systems can compensate for inherent defects in the independent system or instability. Secondly, to achieve constant space ventilated availability, the coupled systems are capable of enhancing ventilation under various climatic scenarios through thermal management (temperature regulation and heat storage) to reduce undesirable temperature fluctuation. Thirdly, once one of the combined systems fails to operate properly under adverse climatic conditions, the other system can continue working, temporarily replacing the failed system to fulfill the thermal requirement. Alternatively, provide the driving force needed by another system to create conditions conducive to operation. These several aspects were further elaborated on in the following several paragraphs.

Firstly, the combined natural ventilation systems can offer advantages that cannot be achieved through a single system. For example, a single system cannot guarantee a stable performance due to uncertain climate, so the combined form that can provide enhancement of ventilation performance under a stable heat exchange becomes a practical strategy. Employing pre-cooling/pre-heating air from the earth-to-air-heat exchanger (shown in Fig. 1.9) is an innovative application of mix-mode ventilation to improve thermal and ventilation performances for space cooling/heating [67]. An earth-air heat exchanger (EAHE) is a technique to circulate the air in the building employing pipes buried at a particular depth underground, utilizing soil beneath the ground as a heat sink to offer an undisturbed constant temperature source [64, 68]. Because of the passive nature of the operation, the implementation of EAHE is still primarily affected by surrounding conditions, such as ambient air temperature, groundwater level, soil type, and outdoor relative humidity. When the soil is thermally saturated, the energy transfer between the soil and the air does not occur, and the temperature gradient between the soil and the air is approaching zero [69].

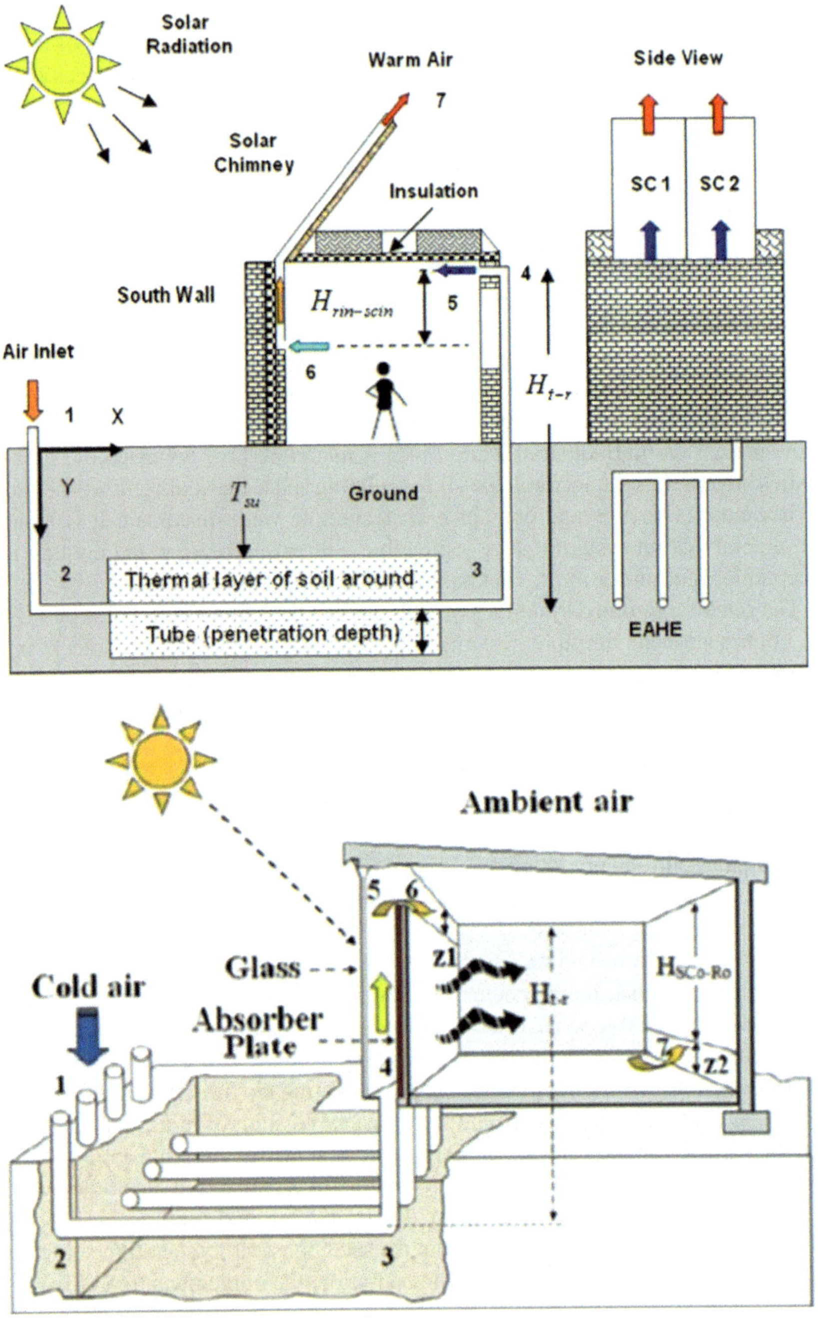

Fig. 1.9 Diagram of the combined earth-air heat exchanger and solar chimney/roof solar chimney [64, 70, 72]

The traditional EAHE is driven by a fan, contrary to the original intention of low-energy building design. Alternatively, the idea of integrating EAHE with buoyancy-based ventilation systems could be the solution to eliminate the use of fans and reduce energy consumption and peak cooling demand in the hot season. Maerefat and Haghighi [64] proved that the proposed EAHE-SC system could achieve thermal comfort in harsh environmental conditions by making an appropriate modification to the system. A schema of the system is illustrated in Fig. 1.9. The heat source for the stack effect is mainly provided by the solar chimney, while the EAHE system relies on the soil to provide the cooling effect. Li et al. [70] verified that the significant cooling capacity of the coupling system could make the indoor temperature 10–13 °C lower than the outdoor temperature. Wei et al. [67] developed a quantitative model of EAHE and the buoyancy generated by the heat source inside the space, considering the nonlinear coupling relationship of the combined effect.

Additionally, coupling wind-based ventilation and EAHE can provide the driving force for airflow in EAHE. Benhammou et al. [71] developed a mathematical model to investigate the performance of a windcatcher coupled with the EAHE and evaluated the thermal performance of the integrating system in a hot and humid climate. Alternatively, there is a view that the system is unsuitable for residential buildings due to the demand for space and the high price.

Secondly, temperature stability is a prerequisite for providing continuous ventilation. There is an increase in the application of the heat recovery ventilator to reduce the heating and cooling demands. A passive ventilation system with a deep cavity and thermal mass structure provides ideal storage space for the installation of heat transfer devices (see Fig. 1.10). The heat emitted by the building can be used as a heat source or radiator, depending on the ambient circumstance. Heat recovery ventilation (HRV) devices, such as heat pipes [73], phase-change material (PCM), rotary thermal wheels, and fixed plate heat exchangers [74, 75], are integrated components in enhancing ventilation performance. Some studies have reviewed existing heat recovery equipment to explore the potential and important considerations of incorporating heat recovery devices into passive ventilation systems [74, 75].

Sensible and latent heat are two classifications of thermal energy that could be continuously recovered by heat recovery devices. Sensible heat refers to the heat of dry air. The heat recovered by latent heat comes from the moisture in the airflow. Devices capable of recovering sensible and latent heat are preferable for integrating into the passive ventilation system. Heat pipes (see Fig. 1.10) and rotary thermal wheels are considered the most promising integrated technology because of excellent thermal efficiency and low-pressure loss throughout the heat recovery equipment [75]. In order to maintain a sufficient airflow rate, a low-pressure loss is expected. Besides, several disadvantages limit the applications of heat recovery. Take rotary thermal wheels for an instant, and they are susceptible to short-circuiting air. In this case, air flowing in one direction will be recirculated in an unexpected direction.

Thirdly, a large number of studies researched phase-change materials (PCMs) employed in passive ventilation systems as it is readily incorporated into the building. PCM can be integrated either into heat modulation or amortization technique (i.e., modify heat gains) or heat dissipation technique (i.e., remove internal heat). Geetha et al. summarized and discussed the applications of passive cooling techniques with and without thermal storage components and collected relevant performance reports for each technique [2]. The thermal mass containing PCMs applies as thermal storage applications are opposed to instantaneous [76]. During the phase transition, keeping the PCM temperature relatively constant is better. Coupling thermal mass and PCM makes keeping the space at the desired uniform temperature worthwhile and significantly reduces the cooling load in milder weather conditions, especially when PCM is combined with night ventilation [77, 78].

Since solar chimney, Trombe wall, and windcatcher are attached to the building envelopes structurally, PCM can entirely replace the thick concrete external wall due to its superior energy storage ability and ease of compatibility with most building elements. Quite a few studies have investigated the consolidation of PCM in the absorb/storage wall of a solar chimney [79–87] or a Trombe wall [88–99] for building ventilation applications. Waqas and Din [100] provided a review of the passive cooling building using PCM as cold storage. Use the cool air to solidify the PCMs and extract the stored coolness during the hot day. It has been claimed that vital

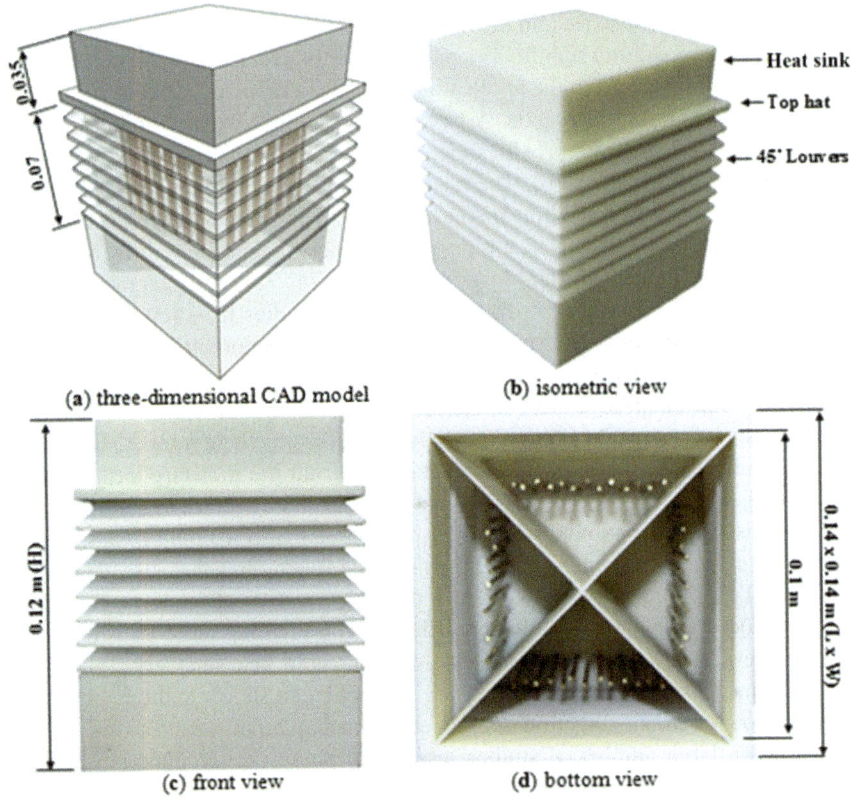

Fig. 1.10 Cooling wind tower/catcher incorporating heat pipes [103, 104]

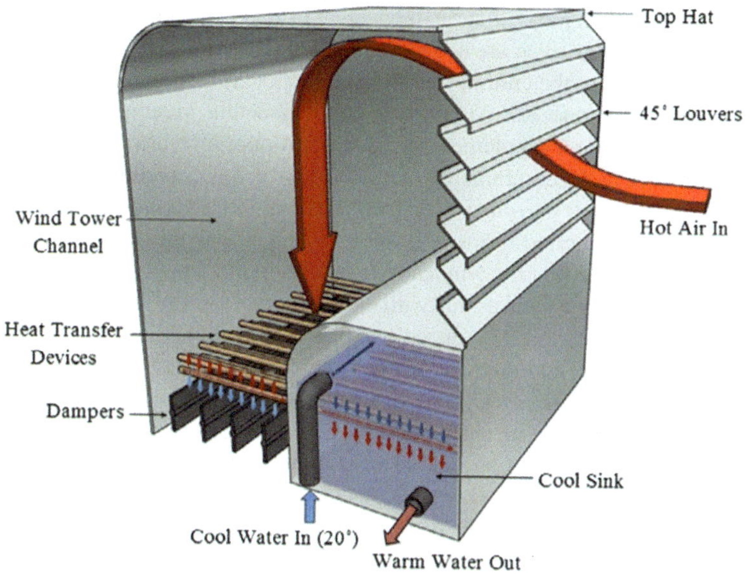

Fig. 1.10 (continued)

challenges and impacts facing the passive system's design provide a wide range of PCMs. Vargas-Lopez et al. [86] presented a wide range of transient mathematical models of SCs with/without PCMs. The limitations and advantages of each model were explored. It further derived a new theoretical model for a double-channel SC with PCMs.

Fluctuations in solar radiation yield solar chimney instability. In the absence of insolation, the thermal efficiency of the solar chimney will decrease sharply. At this time, PCMs as thermal storage mediums will provide heat supplement and heat exchange for the system [79]. The evidence revealed that a solar chimney coupling with PCM could extend the ventilation hours without insolation, notably contributing to ventilation effort during the night time. Calautit et al. [101] obtained a temperature reduction of 9–12 °C achieved by wind tower with a heat transfer device in hot and dry climatic conditions. Xaman et al. [102] performed a conjugate heat transfer analysis under transient conditions for a solar chimney with three different absorbing materials (i.e., concrete, PCM, and copper plate) on the warmest day in Madrid, considering the gains and losses of convection and radiation. The transient numerical simulation proved that the mass flow rates and thermal efficiency values of the solar chimney with a PCM are higher than concrete but lower than a copper plate.

Fourthly, to cope with the inadequacy of the stand-alone system in natural heating/cooling, a benefit of coupling systems is that when the operating capacity of one system is restricted, the other system can still operate normally. Wind-buoyancy ventilation is a strategy adopted to obtain the desired level of thermal comfort and constant naturally ventilated operation [105]. Since relying on a single chimney effect cannot provide sufficient ventilation, the weight of wind ventilation increases. Wind-driven and buoyancy-driven are not mutually exclusive in some cases, and the wind can even contribute to the chimney effect. When there is insufficient solar radiation, the exhaust airflow rate can be supplemented by a fan. Meanwhile, wind-driven devices could provide a supplementary ventilation rate.

The efficiency of the combination of the two driven forces is subject to the interaction between wind and buoyancy effect, which results from many variables, such as flow patterns, incorporated forms, and opening dimension and position. The studies [27, 28, 106] outlined that the windcatcher can be coupled with evaporative cooling systems, solar chimneys, and a courtyard (see Figs. 1.11a, c and 1.12). Abdallah et al. [107] presented a coupled system with a solar roof chimney and a wind tower, as shown in Fig. 1.11c. The integrated system achieved a volume flow rate of 414 m^3/h and a decrease of 10–11.5 °C at room temperature. Besides, a traditional Trombe wall may lose much heat at night due to its low thermal resistance. It might also act as a heat source under hot atmospheric conditions and cause overheating in a well-insulated building [108]. Rabani and Kalantar [109] proposed a newly designed coupling system of the Trombe wall and solar chimney with a water spraying system, as shown in Fig. 1.11d, to enhance cooling capacity. It appraised the system's cooling capacity numerically under the desert climate. It explored that the mass flow of 10 l/h and ACH approaching eight could be achieved by the combined system. It is also proved that the change in outdoor relative humidity just slightly affects the air velocity and ACH. As the temperature difference between the indoor and the cavity decreases, the average air velocity and ACH in the channel will decline.

Meanwhile, the thermal effectiveness of the system can be improved by roughly 30% by a water spraying system. Ahmed et al. [110] reviewed the ventilation performance of single-sided and cross-ventilation, windcatchers, and solar chimneys in a warm climate environment and evaluated thermal comfort and IAQ. The study proposed that combining SC or windcatchers with water evaporation cooling is the best solution to achieve the cooling effect and ventilation efficiency in hot and dry climates.

Nowadays, the concept of a solar-wind tower system employed in a naturally ventilated building has been studied by various researchers numerically and experimentally, as seen in Fig. 1.11a and e [111, 112]. By thermal gradients and incident wind, the enhancement of flow convection can be produced. The glassing cover of a solar chimney (see Fig. 1.11e) is heated by solar radiation and warms the air across the channel. After that, warm air rises and exhausts from outlets and draws fresh air through the openings [31]. Some researchers argued that the application of a solar chimney integrated with a windcatcher might be the solution when there is no desired wind [113]. In other words, the thermal efficiency of a solar chimney is comparatively higher when encountering inadequate incident wind [114]. A wind tower could generate a mass flow rate roughly of 0.75 kg/s at a wind speed of 1.0 m/s, while the solar chimney-assisted system was capable of producing up to a 1.4 kg/s airflow rate at 700 W/m^2 solar intensity [112] and a reduction of air temperature up to 15 °C for the combined system [111]. Bansal et al. [112] developed a steady-state mathematical model of an integrated system with a solar chimney and a wind

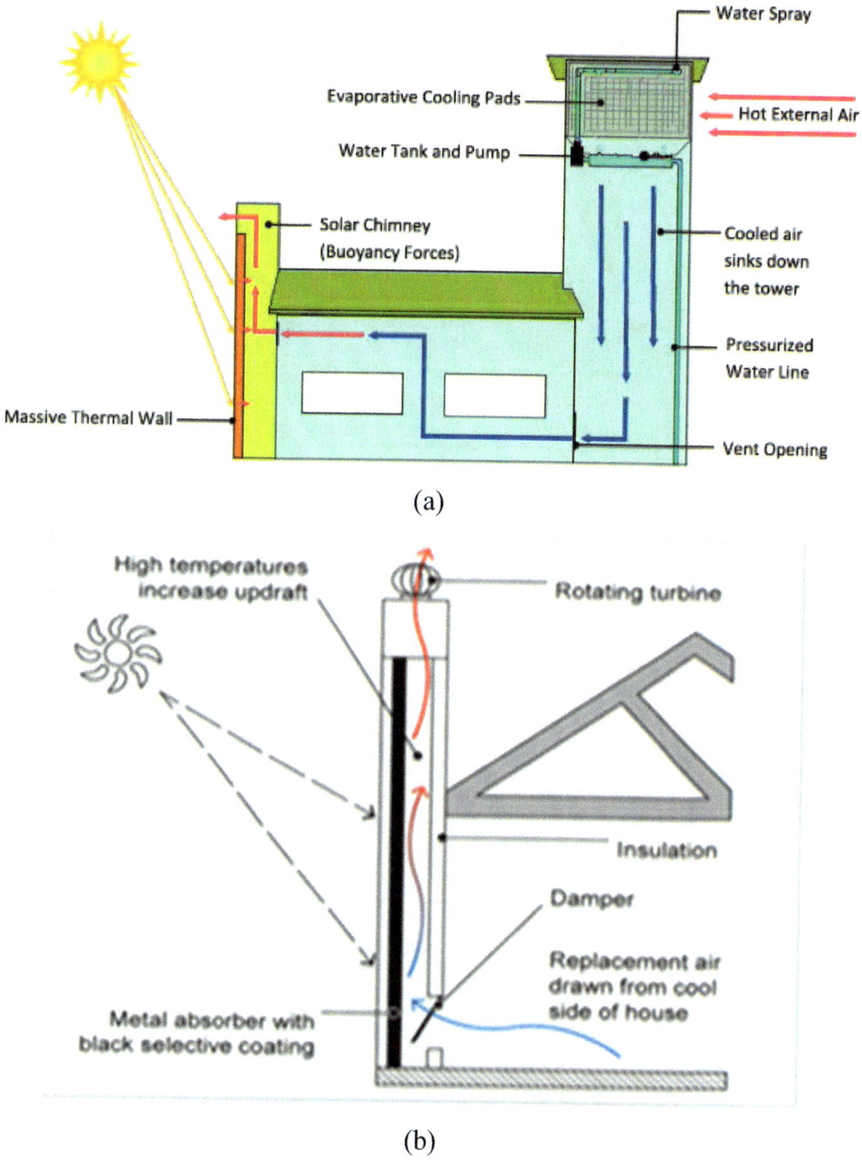

Fig. 1.11 Diagrams of a coupled system with different natural ventilation systems: **a** conceptual design of a wind tower integrated with a solar chimney [27, 111]; **b** combination of a solar chimney and a turbine ventilator (*Source* Green Builder Solar Guidelines (Residential)); **c** solar roof chimney integrated with an evaporative cooling wind tower [107]; **d** schema of a Trombe wall coupled with and a solar chimney and a water spraying system [109]; and **e** wind tower integrated with solar chimney [27]

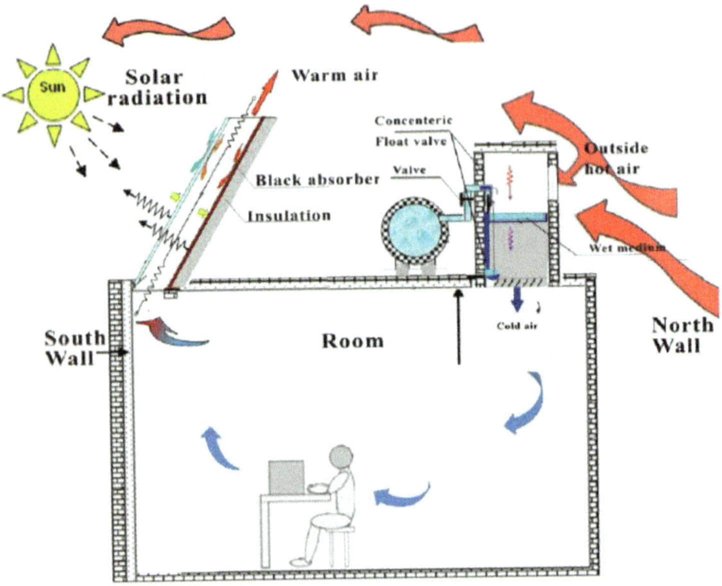

(c)

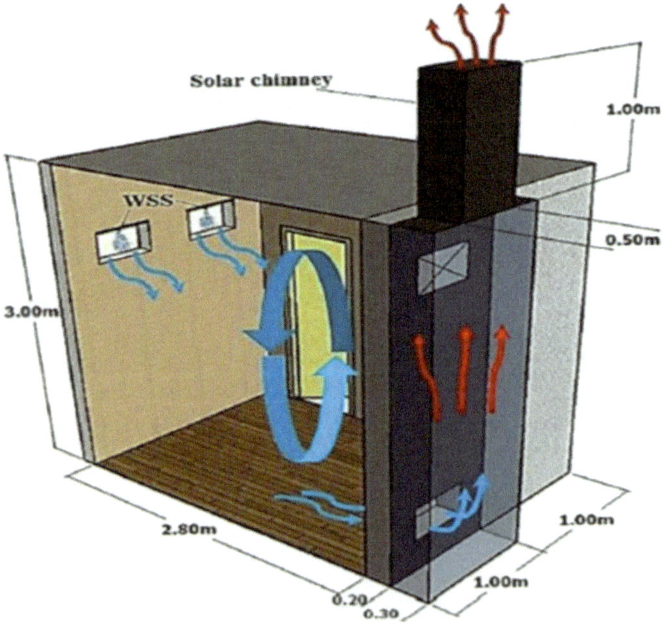

(d)

Fig. 1.11 (continued)

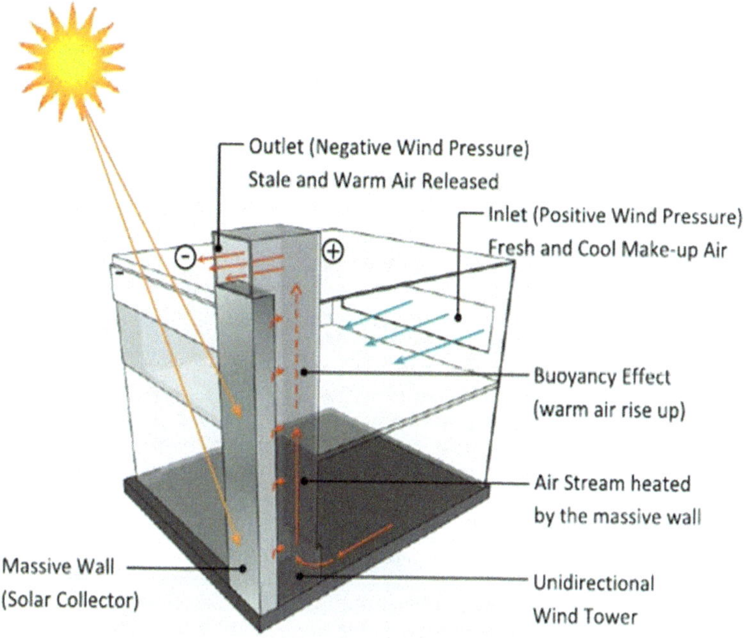

(e)

Fig. 1.11 (continued)

tower to ventilate a three-storey building. An individual wind tower generated only 20 ACH compared to 60 ACH produced by the integrated system. Hughes et al. [27] provided a wide-ranging review of wind tower evolution and evaluated the potential of the combinations. Nouanégué et al. [113] investigated solar-wind tower systems numerically. The impacts of influential geometrical parameters on the ventilation and thermal performance of the solar chimney were identified.

The combined structure of the wind tower-atria/courtyard is a traditional architectural style in a hot and dry area, as illustrated in Fig. 1.12a. The air in the courtyard was heated by solar radiation during the day, and the air movement occurred due to the stack effect, resulting in air circulation between adjoining spaces. In contrast, the courtyard is operated as a heat sink that washes away the stale warm air in the room and fulfills coolness to the adjacent spaces during the night time [115]. Since atria are a long-standing ventilation system, there is much research on adding additional systems to its structure. The combination of the atrium with a solar chimney and a DSF can benefit from intensive solar radiation, as seen in Fig. 1.12b. The interactive support between components could address the insufficiency of the buoyancy force [41].

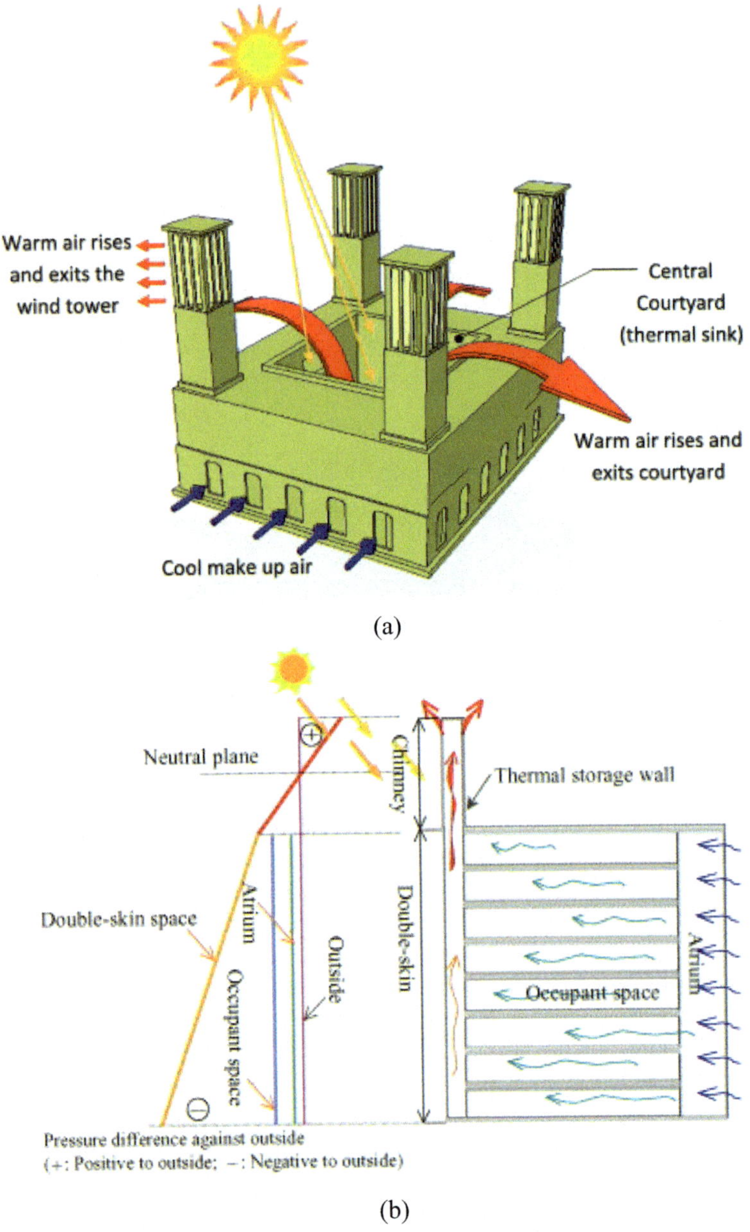

Fig. 1.12 **a** Schematic of integrated wind towers and a courtyard [27]; **b** diagram of an atrium integrated double-skin façade with solar chimney [41]

Similarly, the research on solar chimneys has never stopped. In addition to the related research mentioned in Sects. 1.4 and 1.5, Jafari and Poshtiri [116] established a solar system composed of a solar-driven adsorption chiller, integrating a solar chimney and a cooling channel. The proposed system is simulated analytically, and internal temperature, relative humidity, and ACH are examined. The effects of different contributing parameters on air temperature and ACH were also evaluated.

Finally, combining building design optimization strategies with passive design can provide a more comprehensive and valuable energy-saving scheme for the conceptual design and hypothesis stage. Stevanović [117, 118] extensively reviewed the previous studies on passive solar design strategies based on numerical simulation. The review also introduced the optimization methods available and software packages employed widely, focusing on the building forms, opaque envelope components, glazing types, and shading devices.

1.4.2 Problems and Improvements

Since constraints vary according to the diversity of the modeling conditions and controlled experiments under various climatic conditions, it is challenging to draw comprehensive and in-depth conclusions about integrating systems performance. Consequently, it is intricate to rely on previous research achievements to summarize objectively and practically in detail.

According to the studies mentioned above, most of the combined systems exist in systems based on the principle of buoyancy-driven, while only a small amount of studies dealt with the cooperation of wind-driven and buoyancy-driven due to the complexity of the airflow mechanism. The effects of wind on airflows, especially on the coupling of natural ventilation systems, are remarkable. Yusoff et al. [119] revealed that the influence of wind on the airflow rate might be favorable or unfavorable, depending on the location of the inlet/outlet vents. Simultaneously, the airflow path, ventilation rate, and temperature change are affected by the inherent characteristics of the system itself and the uncertainty of natural conditions.

Although combined passive systems are widely regarded as an effective strategy for sustainable buildings, there is currently no general criterion to summarize and rank them in line with their performance. It is then necessary and critical to conduct more experiments toward coupling natural ventilation systems to solve commercialized disputes in their implementations [29]. The strengths of natural ventilation come from the capability of providing a desired air change rate economically with a straightforward structure. Despite the existence of significant fluctuations in the air change rate, buildings with thorough designs and proper operation can realize the desired ventilation and thermal performance by natural forces under ideal climate conditions.

Additionally, thermal bridge breaking of glazing materials and condensation phenomena may occur in either cold seasons or hot and humid climates. Since natural ventilation systems lack dehumidification and temperature control, these risks may compromise the durability and longevity of buildings. Since the natural ventilation system is mainly driven by wind and thermal buoyancy, fire protection issues in most NV systems have not been resolved. Only a few studies have contributed to early smoke exhaustion [155–158]. For example, Cheng et al. [156] determined the typical factors affecting natural ventilation and fire safety performance based on a solar chimney. A world-first solar chimney was also implemented in a real practice considering energy conservation and fire safety in Melbourne of Australia [159]. Although these studies have confirmed the viability of solar chimneys in energy-saving and fire safety, more studies are critically needed to prepare them for practical usage in real buildings regarding their dual functions. Yang et al. [158] established a quantitative model to study the convection between a vertical shaft and adjacent spaces via parametric studies, especially when the heated current flows from the spaces into the cavity with a lower temperature. The developed model can predict the vertical distribution of temperature and pressure, mass flow rate, NPL position, and the critical conditions for the transition from single flow to two-way flow. Sensitivity studies were used to analyze a series of parameters that affect the buoyancy effect, such as chimney configurations, the physical characteristics of the air, and the convective heat transfer coefficient.

1.4.3 Recommendations and Implementations

In many studies, ventilation enhancement methods have been dedicated to analyzing the performance of natural ventilation systems numerically and experimentally. Although previous research has focused on the functional evaluation and optimization design of single systems, there are few papers on the centralized classification of passive ventilation systems and the comprehensive review of existing combined systems. In order to achieve a comprehensive natural ventilation survey and advance the exploration of combined systems, this study records in detail the various attempts made by researchers in natural ventilation over the past 20 years, especially the combination of passive systems and their performance.

1.4.3.1 Research Progresses on Combining Natural Ventilation Systems

A list of natural ventilation technologies and their combination in a single building were reviewed and analyzed, together with their applicability, advantages, and disadvantages in the previous sections. Figure 1.12 intuitively shows the share of different combination systems in the research over the years. The following observations can be derived and concluded after the review study. The combination of buoyancy and wind-driven ventilation is the most common method—especially the combination of solar chimney and wind tower/catcher, occupying 14.44%. The Trombe wall-HRV combination, which has the same share, has also been numerically simulated and visualized test by many researchers. The secondary research areas are similarly concentrated on combining the HRV system and ventilation facilities with thermal mass, namely SC-HRV and wind tower/catcher-HRV, respectively, occupying about 13%. The third research area is related to nocturnal cooling applications. The radiation effect of day and night can form a natural temperature difference between inside and outside, which can be used by almost any natural ventilation system. Meanwhile, night ventilation is a relatively important sector in the field of natural ventilation.

The ancient ventilation structures, such as the combination of the atrium and wind towers/catchers, which are still used today, constitute the fourth research area. With its excellent thermal insulation and transparent appearance, the double-skin façade has gradually become a new research focus. It is combined with the solar chimney or the wind tower/catcher system, which also has an intermediate cavity, to enhance the buoyancy effect and promote the airflow in the cavity. This combination not only promotes building ventilation but also reduces cooling and heating demand. The remaining system combination forms are sporadic due to the small volume of the ventilation facilities and their restricted ventilation capacity.

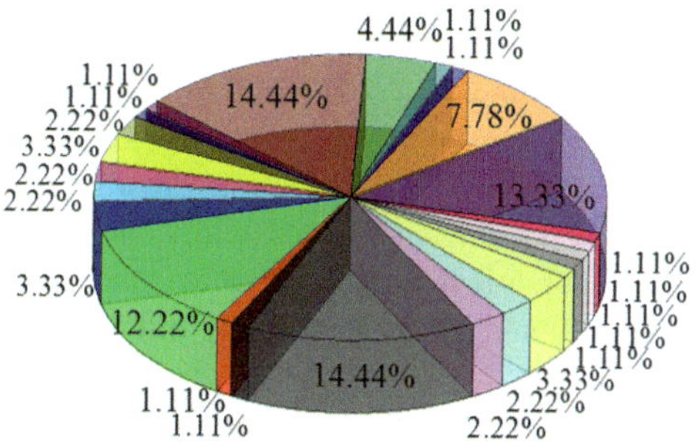

Legends table

Combinations	TW	SC	DSF	A/C	PV	WT/C	F	WW	WS	RC/VT	RV	HRV	EAHE	NC
TW		SC-TW	NA	NA	NA	NA	NA	NA	NA	NA	NA	TW-HRV	NA	NA
SC			SC-DSF	SC-A/C	SC-PV	SC-WT/C	SC-F	NA	SC-WS	NA	NA	SC-HRV	SC-EAHE	SC-NC
DSF				DSF-A/C	DSF-PV	DSF-WT/C	NA	NA	NA	NA	NA	DSF-HRV	NA	NA
A/C					NA	A/C-WT/C	NA	NA	NA	NA	A/C-RV	NA	NA	A/C-NC
PV						NA	NA	NA	NA	NA	NA	NA	NA	NA
WT/C							NA	WT/C -WW	WT/C-WS	WT/C-RC/VT	WT/C-RV	WT/C-HRV	WT/C-EAHE	WT/C-NC
F								NA	NA	NA	NA	NA	NA	NA
WW									NA	NA	NA	NA	NA	NA
WS										NA	NA	NA	NA	NA
RC/VT											NA	NA	NA	NA
RV												NA	NA	NA

Fig. 1.13 Research share of the cross-combination of natural ventilation systems (TW—Trombe wall; SC—solar chimney; DSF—double-skin facade; A/C—atria and courtyard; PV—solar walls with the photovoltaic module; WT—wind tower; WC—windcatcher; F—fenestration (single-sided and cross-ventilation); WW—wing walls; WS—wind cowls/scoops; RC—rotating wind cowl; VT—turbine ventilation; RV—exhaust cowls/roof vents/roof cowls; HRV—heat recovery ventilation devices; HID—heat transfer device; EAHE—earth-air heat exchanger; and NC—nocturnal cooling applications)

1.4.3.2 Current Challenges and Future Works

The following suggestions can be provided by reviewing and summarizing the combined mode of the passive ventilation systems above. Due to its complexity and lack of theoretical models, the majority of previous studies are without detailed parametrical results or merely focus on the validation of analytical models in corresponding experiments. Some studies examined a stand-alone ventilation mode's ventilation capacity and thermal performance under specific climatic conditions [160, 161]. The primary considerations in the improvement of the performance of the natural ventilation approach include introducing different configurations of systems (e.g., configurations and geometries of systems, stack height, the inclination angle of a solar collector and cavity depth) and structural dimensions (e.g., building mass, window-to-wall ratio), as well as the parameters affected by external and internal variables [162], such as the solar absorptance of the thermal mass [32], heat transmittance and solar reflectance of glazing materials [42], and ambient conditions (e.g., solar insolation, wind velocity, and direction) [32, 162]. For example, many mathematical models have been proposed to predict the performance of a solar chimney based on the above variables [65, 163–167]. Some studies also focus on the correlation between the calculated airflow rate or thermal performance and the variables used based on the parametric analysis [13, 28, 168]. It is suggested to develop the parametric analysis by considering more contributing variables in the future work.

Secondly, according to the leading indicators (temperature and volume airflow rate), buoyancy-driven ventilation is mainly affected by solar radiance and ambient temperature, while wind speed and direction strongly influence wind-driven ventilation. The stack performance predictions are relatively reliable at the design stage [169]. Conversely, it is difficult to predict and control the potential performance of wind-driven ventilation due to its stochastic characteristics [170]. Although wind-induced force can generate a more considerable airflow rate inside the building than the buoyancy effect, air flow could remain constant with proper buoyancy design. The stack effect tends to be less intense. Thus, the wind-induced force requires compromise with buoyancy force for better collaboration in designing and integrating passive systems, especially for districts with low internal and external temperature differences [171]. When combining such systems, a thoughtful investigation is required to avert the unpredictable delivery of air flow.

Thirdly, a thoughtful attempt was made at design considerations of the combinations over the year. Most of those previous studies were found to concentrate on one-storey buildings. With the growth of the population, constructors tend to build multi-storey buildings instead. Therefore, the solar chimney or DSF is an attractive solution for a multi-storey structure. Besides, the components with the function of temperature regulation and exchange are well integrated into existing ventilation systems. Studies include using EAHE-based wind towers/solar chimneys and PCM-based Trombe wall/solar chimney/wind towers to improve ventilation performance and thermal performance. The proposed combination has been developed numerically and examined experimentally under a controlled environment. In 2D ventilated rooms in contact with cold external environments, the heat transfer of natural convection and surface radiation has been studied numerically [160]. The authors explored the noticeable dependence of thermal performance on the geometrical features of solar chimneys and external variations [160]. Due to the increase in building volume, multiple combination systems will become a development trend in energy-efficient buildings. Salari et al. [126] presented a numerical study of a compound rooftop solar chimney with the PV module and PCM and compared the performance of the proposed system with a conventional SC, an independent PV module, an SC-PV system, and an SC-PCM system. The proposed system shows superiority in power generation under a subtropical climate. In order to achieve more significant building energy savings while meeting the building's internal cooling and ventilation requirements, Elghamry and Hassan [136] also used the power generation capacity of PV to propose and examine a combination of three systems called a geothermal tube-chimney-PV system. The experimental results show that the new system can increase the room temperature by 6.4 °C and generate a forced airflow at 0.0184 m^3/s.

Moreover, as the influencing factors of the combined system increase significantly, it is necessary to carefully consider as many influencing factors as possible when predicting system performance instead of blindly combining according to the performance advantages of each system. Taking SC-HRV for example, the performance of the coupling of a solar chimney and heat recovery devices was investigated in [73]. Using a solar chimney with different glazing materials to warm the air in the cavity, the ventilation rate increases under a higher cavity air temperature. When the model added heat pipes to try to recover the thermal energy in the air, it was found that the buoyancy effect was reduced, thereby limiting the ventilation rate. It also suggested that the buoyancy effect drives passive ventilation often requires external wind assistance to drive the airflow to the required volume flow rate.

Furthermore, studies on the coupling between different passive ventilation systems are still insufficient. The effort in the future works includes the need for parameterization studies. The influences of different kinds of parameters affecting the performance of natural ventilation systems on combined passive systems could be further investigated. Due to the limitations of research methods and experimental conditions, although some combination modes' effectiveness has been practiced, obtaining a universal combination mode of passive ventilation approaches and recognized accurate prediction methods has not been possible. Figure 1.12 and Table show that the leading integrated natural ventilation systems existing are solar chimney-wind tower systems, wind tower-courtyards/atrium systems, and atrium-SC-DSF systems. The airflow mechanisms of integrating techniques, along with their complementary capabilities and applicability, are discussed in the relevant literature. However, there are significant differences between the predicted results and experimental data of passive ventilation systems mentioned in the selected literature summarized in Table 1.1.

Additionally, another direction that deserves attention and effort is the coupling of the thermal mass of the building itself with buoyancy-driven ventilation. Yang and

Guo [172] have proved that the ventilation rate is coupled with the indoor air temperature. Such interaction, in turn, will change the internal temperature and ventilation rate in thermal pressure ventilation.

It is worth mentioning that the applications of integrating the evaporative cooling device with natural ventilation systems to improve thermal conditions and reduce the energy demand of ventilation and cooling [28, 29, 140, 173–175] exist in numerous case studies. Since most systems use water as a medium, it is not sustainable for water-scarce areas. Besides, it was barely noticeable to see some ventilation devices, such as wind scoops, turbine ventilators, and roof cowls, applied to the integrating system. Even though they are capable of obtaining the desired IAQ for a single room, the combined applications may not be practical in high-mass residential buildings due to their limited ventilation capacity. Such widgets are often incorporated as appendages into more extensive systems driven by the wind, like wind cowls/scoops, in conjunction with wind towers [176].

Last but not least, to obtain a green building and optimize the building's structure and the equipment that adjusts the indoor environment, sustainability, and the low-cost design process also receive research attention. Using refined building information modeling (BIM) models to simulate and analyze buildings is one of the most promising methods in the architecture, engineering, and construction (AEC) industry [177]. BIM performs pre-operation and physical simulation of buildings by integrating different simulation software, which can comprehensively analyze architectural lighting (i.e., Ecotect), wind environment (i.e., Ansys Fluent), and structural loads (i.e., Abaqus) to form an intuitive three-dimensional architectural model. BIM makes the designed structure more intuitive. After continuous simulation verification, the architectural plan's optimal design is outlined using BIM software to evaluate green buildings and BIM functions in practical applications. Combine the three-dimensional collaborative design under the BIM with the investigation of green buildings, make full use of the richness of BIM information and the foresight of physical simulation of green buildings, provide a scientific basis for architectural design, and provide designers with solid technical support [178, 179].

1.5 A Focus on Buoyancy Only: Rationale

The solar chimney has gained popularity as an ideal component of passive ventilation systems and a supplementary or alternative strategy for mechanical ventilation due to its superior structural qualities and economic and environmental advantages. The wind environment shows a significant impact on solar chimney performance. However, we did not consider the impacts of the wind due to two reasons: (1) Afonso and Oliveira [180] mentioned that 'the design of a solar chimney can be done without considering the wind effect' as without considering wind can underestimate the actual ventilation rates conservatively; and (2) as the wind environment is different by the location, the impacts of wind environment were ignored; otherwise, it will distract the analysis that the obtained results cannot provide a benchmark result to be adopted in different places. The benchmark analysis considering the wind effect will be carried out in the future works.

1.6 Solar-Assisted Passive Ventilation

Existing passive ventilation systems that use solar energy as the main driving force to enhance natural ventilation include solar chimneys (SC), atrium, naturally ventilated double-skin façade (NDSF), and Trombe walls. Solar-induced approaches employ direct or indirect solar energy to capture solar radiation and convert energy into usable heat [7]. Solar radiation intensity is one of the external environmental elements that may directly increase buoyancy-driven ventilation efficiency without considering the external wind influence.

The working mechanism of conventional passive solar designs is identical. The energy of solar radiation penetrates through the outer glass to the inner absorber through natural convection and heat transfer that causes a temperature gradient in the vertical air channel attached to the building envelope, thereby producing thermal buoyancy to promote an upward airflow in the air layer. The occupant's room has a vent connected to the air layer. The pressure difference generated by air density deference at vents drives the air movement from the adjacent spaces to the air channel to develop indoor air circulation or discharge the stagnant indoor air through the designed export. More details are introduced in Chap. 3.

References

1. Pérez-Lombard L, Ortiz J, Pout C (2008) A review on buildings energy consumption information. Energy Build 40(3):394–398
2. Geetha N et al (2012) Passive cooling methods for energy efficient buildings with and without thermal energy storage—a review 29(2):913–946
3. Zhang H et al (2021) A critical review of combined natural ventilation techniques in sustainable buildings. Renew Sustain Energy Rev 141:110795
4. Santamouris M, Kolokotsa D (2013) Passive cooling dissipation techniques for buildings and other structures: the state of the art. Energy Build 57:74–94
5. Chartier Y, Pessoa-Silva C (2009) Natural ventilation for infection control in health-care settings. World Health Organization
6. Ahmed T, Kumar P, Mottet L (2021) Natural ventilation in warm climates: the challenges of thermal comfort, heatwave resilience and indoor air quality. Renew Sustain Energy Rev 138:110669
7. Quesada G et al (2012) A comprehensive review of solar facades. Opaque solar facades. Renew Sustain Energy Rev 16(5):2820–2832
8. Tao Y et al (2021) Ventilation performance of a naturally ventilated double skin façade with low-e glazing. Energy 120706
9. Etheridge DW, Sandberg M (1996) Building ventilation: theory and measurement, vol 50. Wiley Chichester
10. Chan H-Y, Riffat SB, Zhu J (2010) Review of passive solar heating and cooling technologies. Renew Sustain Energy Rev 14(2):781–789

11. Awbi H (2010) Basic concepts for natural ventilation of buildings. In: CIBSE BSG seminar: natural and mixed-mode ventilation modelling, University of Reading
12. Pomponi F et al (2016) Energy performance of double-skin Façades in temperate climates: a systematic review and meta-analysis 54:1525–1536
13. Al Assaad D, Ghali K, Ghaddar N (2019) Effect of flow disturbance induced by walking on the performance of personalized ventilation coupled with mixing ventilation. Build Environ 160
14. Fernandez-Aguera J et al (2019) Thermal comfort and indoor air quality in low-income housing in Spain: the influence of airtightness and occupant behaviour. Energy Build 199:102–114
15. Oropeza-Perez I, Ostergaard PA (2018) Active and passive cooling methods for dwellings: a review. Renew Sustain Energy Rev 82:531–544
16. Chen T et al (2020) Study of flow characteristics in tunnels induced by canyon wind. J Wind Eng Ind Aerodyn 202:1–12
17. Wang QY et al (2020) Numerical simulation of two-way fluid-structure interaction of wind loading on buildings. J Chin Inst Eng 43:225–240
18. Santamouris M, Wouters P (2006) Building ventilation: the state of the art. Routledge
19. Liddament MW, Infiltration AIDLÉA (1996) A guide to energy efficient ventilation. Air Infiltration and Ventilation Centre Coventry
20. Ashrae (2005) 2005 ASHRAE handbook: fundamentals. ASHRAE
21. Kleiven T (2003) Natural ventilation in buildings: architectural concepts, consequences and possibilities. Institutt for byggekunst, historie og teknologi
22. Handbook AJASOH (2001) Refrigerating and A. Air-conditioning engineers, GA, USA, fundamentals (Chapter 6)
23. Mao JC, Yang WW, Gao NP (2015) The transport of gaseous pollutants due to stack and wind effect in high-rise residential buildings. Build Environ 94:543–557
24. Yang T, Clements-Croome DJ (2012) Natural ventilationnatural ventilationin built environmentnatural ventilationin-built environment. In: Meyers RA, (ed) Encyclopedia of sustainability science and technology. Springer New York: New York, NY. pp 6865–6896
25. Gan GH (2000) Effective depth of fresh air distribution in rooms with single-sided natural ventilation. Energy Build 31(1):65–73
26. Khan N, Su Y, Riffat SB (2008) A review on wind driven ventilation techniques. Energy Build 40(8):1586–1604
27. Hughes BR, Calautit JK, Ghani SA (2012) The development of commercial wind towers for natural ventilation: a review. Appl Energy 92:606–627
28. Jomehzadeh F et al (2017) A review on windcatcher for passive cooling and natural ventilation in buildings, part 1: indoor air quality and thermal comfort assessment. Renew Sustain Energy Rev 70:736–756
29. Monghasemi N, Vadiee A (2018) A review of solar chimney integrated systems for space heating and cooling application. Renew Sustain Energy Rev 81:2714–2730
30. Zhang GM, Shi L (2018) Improving the performance of solar chimney by addressing the designing factors. IOP Conf Ser: Earth Environ Sci 168:1–6
31. Arce J et al (2009) Experimental study for natural ventilation on a solar chimney. Renew Energy 34(12):2928–2934
32. Lee KH, Strand RK (2009) Enhancement of natural ventilation in buildings using a thermal chimney. Energy Build 41(6):615–621
33. Zhai XQ, Song ZP, Wang RZ (2011) A review for the applications of solar chimneys in buildings. Renew Sustain Energy Rev 15(8):3757–3767
34. Shi L et al (2016) Developing an empirical model for roof solar chimney based on experimental data from various test rig. Build Environ 110:115–128
35. Thirugnanasambandam M, Iniyan S, Goic R (2010) A review of solar thermal technologies☆. Renew Sustain Energy Rev 14(1):312–322
36. Quesada G et al (2012) A comprehensive review of solar facades. Transparent and translucent solar facades. Renew Sustain Energy Rev 16(5):2643–2651

37. Asfour OJNVS (2015) Health implications and I.O.T. environment. Natural ventilation in buildings: an overview, pp 1–26
38. Shameri MA et al (2011) Perspectives of double skin facade systems in buildings and energy saving. Renew Sustain Energy Rev 15(3):1468–1475
39. Hay R, Ostertag CPJB (2018) Life cycle assessment (LCA) of double-skin façade (DSF) system with fiber-reinforced concrete for sustainable and energy-efficient buildings in the tropics. Environment 142:327–341
40. Perez-Grande I, Meseguer J, Alonso GJATE (2005) Influence of glass properties on the performance of double-glazed facades 25(17–18):3163–3175
41. Ding W et al (2005) Natural ventilation performance of a double-skin façade with a solar chimney 37(4):411–418
42. Barbosa S, Ip K (2014) Perspectives of double skin facades for naturally ventilated buildings: a review. Renew Sustain Energy Rev 40:1019–1029
43. Gratia E, De Herde AJE (2007) Greenhouse effect in double-skin facade. Buildings 39(2):199–211
44. Høseggen R et al (2008) Building simulation as an assisting tool in decision making: case study: with or without a double-skin façade? 40(5):821–827
45. Lou WJ et al (2012) Experimental and zonal modeling for wind pressures on double-skin facades of a tall building. Energy Build 54:179–191
46. Pasquay T (2004) Natural ventilation in high-rise buildings with double facades, saving or waste of energy. Energy Build 36(4):381–389
47. Olsen EL, Chen QYJE (2003) Energy consumption and comfort analysis for different low-energy cooling systems in a mild climate. Buildings 35(6):560–571
48. Wang F, Abdullah AHJIJOL-CT (2011) Investigating thermal conditions in a tropic atrium employing CFD and DTM techniques 6(3):171–186
49. Holford JM, Hunt GR (2003) Fundamental atrium design for natural ventilation. Build Environ 38(3):409–426
50. Horan JM, Finn DPJE (2008) Sensitivity of air change rates in a naturally ventilated atrium space subject to variations in external wind speed and direction. Buildings 40(8):1577–1585
51. Taleghani M, Tenpierik M, van den Dobbelsteen A (2014) Energy performance and thermal comfort of courtyard/atrium dwellings in The Netherlands in the light of climate change. Renew Energy 63:486–497
52. Aldawoud A, Clark RJE (2008) Comparative analysis of energy performance between courtyard and atrium in buildings. Buildings 40(3):209–214
53. Chen S et al (2020) A review of internal and external influencing factors on energy efficiency design of buildings. Energy Build 216:1–17
54. Moosavi L et al (2014) Thermal performance of atria: an overview of natural ventilation effective designs. Renew Sustain Energy Rev 34:654–670
55. Hussain S, Oosthuizen PH, Kalendar A (2012) Evaluation of various turbulence models for the prediction of the airflow and temperature distributions in atria. Energy Build 48:18–28
56. Lin YL et al (2018) Development of building thermal load and discomfort degree hour prediction models using data mining approaches. Energies 11:1570–1583
57. Rundle CA et al (2011) Validation of computational fluid dynamics simulations for atria geometries. Build Environ 46(7):1343–1353
58. Gan GH, Riffat SB (2004) CFD modelling of air flow and thermal performance of an atrium integrated with photovoltaics. Build Environ 39(7):735–748
59. Lomas KJ, Cook MJ, Fiala D (2007) Low energy architecture for a severe US climate: design and evaluation of a hybrid ventilation strategy. Energy Build 39(1):32–44
60. Raji B et al (2020) Natural summer ventilation strategies for energy-saving in high-rise buildings: a case study in the Netherlands. Int J Vent 19(1):25–48
61. Yang D, Li P (2015) Natural ventilation of lower-level floors assisted by the mechanical ventilation of upper-level floors via a stack. Energy Build 92:296–305
62. Yang D, Li P (2015) Dimensionless design approach, applicability and energy performance of stack-based hybrid ventilation for multi-story buildings. Energy 93:128–140

63. Amirifard F, Sharif SA, Nasiri F (2019) Application of passive measures for energy conservation in buildings—a review. Advan Build Energy Res 13(2):282–315
64. Maerefat M, Haghighi AP (2010) Passive cooling of buildings by using integrated earth to air heat exchanger and solar chimney. Renew Energy 35(10):2316–2324
65. Mathur J et al (2006) Experimental investigations on solar chimney for room ventilation 80(8):927–935
66. Hussain S, Oosthuizen PH (2012) Numerical investigations of buoyancy-driven natural ventilation in a simple atrium building and its effect on the thermal comfort conditions. Appl Therm Eng 40:358–372
67. Wei HB et al (2018) Coupling of earth-to-air heat exchangers and buoyancy for energy efficient ventilation of buildings considering dynamic thermal behavior and cooling/heating capacity. Energy 147:587–602
68. Khabbaz M et al (2016) Experimental and numerical study of an earth-to-air heat exchanger for air cooling in a residential building in hot semi-arid climate. Energy Build 125:109–121
69. Niu F et al (2015) Investigation on soil thermal saturation and recovery of an earth to air heat exchanger under different operation strategies 77:90–100
70. Li HR et al (2014) Performance of a coupled cooling system with earth-to-air heat exchanger and solar chimney. Renew Energy 62:468–477
71. Benhammou M et al (2015) Performance analysis of an earth-to-air heat exchanger assisted by a wind tower for passive cooling of buildings in arid and hot climate. Energy Convers Manage 91:1–11
72. Haghighi AP, Maerefat M (2015) Design guideline for application of earth-to-air heat exchanger coupled with solar chimney as a natural heating system. Int J Low-Carbon Technol 10(3):294–304
73. Gan G, Riffat SB (1998) A numerical study of solar chimney for natural ventilation of buildings with heat recovery. Appl Therm Eng 18(12):1171–1187
74. O'Connor D, Calautit JKS, Hughes B (2016) A review of heat recovery technology for passive ventilation applications. Renew Sustain Energy Rev 54:1481–1493
75. Mardiana-Idayu A, Riffat SB (2012) Review on heat recovery technologies for building applications. Renew Sustain Energy Rev 16(2):1241–1255
76. Kośny J (2015) PCM-enhanced building components: an application of phase change materials in building envelopes and internal structures. Springer
77. Piselli C et al (2020) Optimal control of natural ventilation as passive cooling strategy for improving the energy performance of building envelope with PCM integration. Renew Energy 162:171–181
78. Prabhakar M et al (2020) Improving the energy efficiency of passive PCM system using controlled natural ventilation. Energy Build 228:110483
79. Li YC, Liu SL (2014) Experimental study on thermal performance of a solar chimney combined with PCM. Appl Energy 114:172–178
80. Kaneko Y et al (2006) Ventilation performance of solar chimney with built-in latent heat storage. In: Proceedings of 10th international conference of thermal energy conference (ECOSTOCK)
81. Amori KE, Mohammed SW (2012) Experimental and numerical studies of solar chimney for natural ventilation in Iraq. Energy Build 47:450–457
82. Li YC, Liu SL (2014) Numerical study on thermal behaviors of a solar chimney incorporated with PCM. Energy Build 80:406–414
83. Liu SL, Li YC (2015) An experimental study on the thermal performance of a solar chimney without and with PCM. Renew Energy 81:338–346
84. Li Y (2013) Thermal performance analysis of a PCM combined solar chimney system for natural ventilation and heating/cooling. Coventry University
85. Dordelly JCF, El Mankibi M, Coillot M (2018) Numerical and experimental investigations of a PCM integrated solar chimney. In: 2018 4th international conference on renewable energies for developing countries (REDEC). IEEE

86. Vargas-López R et al (2019) Mathematical models of solar chimneys with a phase change material for ventilation of buildings: a review using global energy balance. Energy 170:683–708
87. Dordelly JCF et al (2019) Experimental analysis of a PCM integrated solar chimney under laboratory conditions. Sol Energy 188:1332–1348
88. Zhou YK, Yu CW (2019) The year-round thermal performance of a new ventilated Trombe wall integrated with phase change materials in the hot summer and cold winter region of China. Indoor Built Environ 28(2):195–216
89. Li W, Chen W (2019) Numerical analysis on the thermal performance of a novel PCM-encapsulated porous heat storage Trombe-wall system. Sol Energy 188:706–719
90. Liu XH et al (2019) Optimization of a new phase change material integrated photovoltaic/thermal panel with the active cooling technique using Taguchi method. Energies 12(6)
91. Liu XH, Zhou YK, Zhang GQ (2018) Numerical study on cooling performance of a ventilated Trombe wall with phase change materials. Build Simul 11(4):677–694
92. Leang E et al (2017) Numerical study of a composite Trombe solar wall integrating microencapsulated PCM. 122:1009–1014
93. Li SS et al (2019) Numerical study on thermal performance of PCM Trombe wall. Innovative Solutions Energy Transitions 158:2441–2447
94. Kolaitis DI, Founti MA (2017) Solar wall enhanced with phase-change materials: a detailed numerical simulation study. Adv Build Energy Res 11(1):87–103
95. Cabeza LF et al (2007) Use of microencapsulated PCM in concrete walls for energy savings 39(2):113–119
96. Zalewski L et al (2012) Experimental study of small-scale solar wall integrating phase change material. Sol Energy 86(1):208–219
97. Bourdeau LE (1980) Study of two passive solar systems containing phase change materials for thermal storage. Los Alamos Scientific Lab., NM (USA)
98. Onishi J, Soeda H, Mizuno P (2001) Numerical study on a low energy architecture based upon distributed heat storage system. Renew Energy 22(1–3):61–66
99. Khalifa AJN, Abbas EFJE (2009) A comparative performance study of some thermal storage materials used for solar space heating. Buildings 41(4):407–415
100. Waqas A, Din ZUJR (2013) Phase change material (PCM) storage for free cooling of buildings—a review. Renew Sustain Energy Rev 18:607–625
101. Calautit JK, Hughes BR (2014) Integration and application of passive cooling within a wind tower for hot climates. HVAC&R Res 20(7):722–730
102. Xaman J et al (2019) Transient thermal analysis of a solar chimney for buildings with three different types of absorbing materials: copper plate/PCM/concrete wall. Renew Energy 136:139–158
103. Calautit JK, Hughes BR (2016) A passive cooling wind catcher with heat pipe technology: CFD, wind tunnel and field-test analysis. Appl Energy 162:460–471
104. Calautit JK, O'Connor D, Hughes BR (2016) A natural ventilation wind tower with heat pipe heat recovery for cold climates. Renew Energy 87:1088–1104
105. Liu PC, Lin HT, Chou JH (2009) Evaluation of buoyancy-driven ventilation in atrium buildings using computational fluid dynamics and reduced-scale air model. Build Environ 44(9):1970–1979
106. Dehghani-sanij AR, Soltani M, Raahemifar K (2015) A new design of wind tower for passive ventilation in buildings to reduce energy consumption in windy regions. Renew Sustain Energy Rev 42:182–195
107. Abdallah ASH et al (2013) Integration of evaporative cooling technique with solar chimney to improve indoor thermal environment in the New Assiut City. Egypt 4(1):45
108. Stazi F, Mastrucci A, di Perna C (2012) The behaviour of solar walls in residential buildings with different insulation levels: an experimental and numerical study. Energy Build 47:217–229
109. Rabani M et al (2015) Empirical investigation of the cooling performance of a new designed Trombe wall in combination with solar chimney and water spraying system. Energy Build 102:45–57

110. Ahmed T, Kumar P, Mottet L Natural ventilation in warm climates: the challenges of thermal comfort, heatwave resilience and indoor air quality. Renew Sustain Energy Rev 138:110669
111. Kalantar V (2009) Numerical simulation of cooling performance of wind tower (Baud-Geer) in hot and arid region. Renew Energy 34(1):246–254
112. Bansal NK, Mathur R, Bhandari MS (1994) A study of solar chimney assisted wind tower system for natural ventilation in buildings. Build Environ 29(4):495–500
113. Nouanégué H, Alandji L, Bilgen EJRE (2008) Numerical study of solar-wind tower systems for ventilation of dwellings 33(3):434–443
114. Bansal N et al (1993) Solar chimney for enhanced stack ventilation 28(3):373–377
115. Safarzadeh H, Bahadori MJB (2005) Passive cooling effects of courtyards. Environment 40(1):89–104
116. Jafari A, Poshtiri AH (2017) Passive solar cooling of single-storey buildings by an adsorption chiller system combined with a solar chimney. J Clean Prod 141:662–682
117. Stevanovic S (2013) Optimization of passive solar design strategies: a review. Renew Sustain Energy Rev 25:177–196
118. Stevanović S (2013) Optimization of passive solar design strategies: a review. Renew Sustain Energy Rev 25:177–196
119. Yusoff WFM et al (2010) Enhancement of stack ventilation in hot and humid climate using a combination of roof solar collector and vertical stack. Build Environ 45(10):2296–2308
120. Saadatian O et al (2012) Trombe walls: a review of opportunities and challenges in research and development. Renew Sustain Energy Rev 16(8):6340–6351
121. Sung UJ et al (2013) Analysis of reduced cooling load for a multistorey-building incorporating a ventilated double skin facade with a solar chimney channel. Int J Vent 11(4):381–391
122. Preet S et al (2020) Performance evaluation of photovoltaic double-skin facade with forced ventilation in the composite climate. J Build Eng 32:101733
123. Gratia E, De Herde A (2007) Guidelines for improving natural daytime ventilation in an office building with a double-skin facade. Sol Energy 81(4):435–448
124. Priyadarsini R, Cheong KW, Wong NH (2004) Enhancement of natural ventilation in high-rise residential buildings using stack system. Energy Build 36(1):61–71
125. Navarro L et al 2016 Thermal energy storage in building integrated thermal systems: a review. Part 1. Active storage systems 88:526–547
126. Salari A, Ashouri M, Hakkaki-Fard A (2020) On the performance of inclined rooftop solar chimney integrated with photovoltaic module and phase change material: a numerical study. Sol Energy 211:1159–1169
127. Nanda AK, Panigrahi CK (2016) A state-of-the-art review of solar passive building system for heating or cooling purpose. Front Energy 10(3):347–354
128. Bansal N et al (1994) A study of solar chimney assisted wind tower system for natural ventilation in buildings 29(4):495–500
129. Abdallah ASH et al (2014) Parametric investigation of solar chimney with new cooling tower integrated in a single room for New Assiut city. Egypt Clim 5(2–3):92
130. Moosavi L et al (2020) New design for solar chimney with integrated windcatcher for space cooling and ventilation. Build Environ 106785
131. Prajongsan P, Sharples S (2012) Enhancing natural ventilation, thermal comfort and energy savings in high-rise residential buildings in Bangkok through the use of ventilation shafts. Build Environ 50:104–113
132. Abaza HF (2005) An integrated design and control strategy for energy-efficient buildings
133. Zhai XQ, Wang RZ, Dai YJ (2004) Solar chimney combined with underground vent for natural ventilation in energy saving of buildings. Proceedings of the 3rd international symposium on heat transfer enhancement and energy conservation, vols 1 and 2, pp 1117–1123
134. Safari M, Torabi FJEE (2014) Improvement of thermal performance of a solar chimney based on a passive solar heating system with phase-change materials. Systems 2(2):141–154
135. Serageldin AA, Abdelrahman AK, Ookawara S (2018) Parametric study and optimization of a solar chimney passive ventilation system coupled with an earth-to-air heat exchanger. Sustain Energy Technol Assess 30:263–278

136. Elghamry R, Hassan H (2020) An experimental work on the impact of new combinations of solar chimney, photovoltaic and geothermal air tube on building cooling and ventilation. Sol Energy 205:142–153
137. Macias M et al (2006) Application of night cooling concept to social housing design in dry hot climate 38(9):1104–1110
138. AboulNaga M, Abdrabboh SJRE (2000) Improving night ventilation into low-rise buildings in hot-arid climates exploring a combined wall–roof solar chimney 19(1–2):47–54
139. Sharples S, Bensalem RJSE (2001) Airflow in courtyard and atrium buildings in the urban environment: a wind tunnel study 70(3):237–244
140. Edwards B et al (2006) Courtyard housing: past, present and future. Taylor & Francis
141. Bahadori M, Haghighat FJB (1985) Passive cooling in hot, arid regions in developing countries by employing domed roofs and reducing the temperature of internal surfaces. Environment 20(2):103–113
142. Calautit JK et al (2015) CFD and wind tunnel study of the performance of a multi-directional wind tower with heat transfer devices. Clean, Efficient Affordable Energy Sustain Future 75:1692–1697
143. O'Connor D, Calautit JK, Hughes BR (2014) A study of passive ventilation integrated with heat recovery. Energy Build 82:799–811
144. Mardiana A et al (2013) Integrated heat recovery system with wind-catcher for building applications: towards energy-efficient technologies pp 720–7
145. Calautit JK et al (2013) CFD analysis of a heat transfer device integrated wind tower system for hot and dry climate. Appl Energy 112:576–591
146. Calautit JK et al (2013) Comparison between evaporative cooling and a heat pipe assisted thermal loop for a commercial wind tower in hot and dry climatic conditions. Appl Energy 101:740–755
147. Riffat S, Gan GJATE (1998) Determination of effectiveness of heat-pipe heat recovery for naturally-ventilated buildings 18(3–4):121–130
148. Ghadiri MH et al (2011) The effect of tower height in square plan wind catcher on its thermal behavior 5(9):381–385
149. Kirk S (2002) Wind driven top-down natural ventilation and heat gain removal for nondomestic buildings-ground and first floor temperature profiles recorded in relation to external air temperature from the 20/07/02 to 14/08/02 at the Tonbridge and Malling council offices in Kings Hill. Proceedings of the 2002 conference for the engineering doctorate in environmental technology
150. Jones B et al (2007) Air quality measured in a classroom served by roof mounted natural ventilation windcatchers. Heliotopos Conferences Limited
151. Ji Z et al (2012) Performance evaluation and energy saving potential of windcatcher natural ventilation systems in China 1(2):84–95
152. Hughes BR, Chaudhry HN, Ghani SA (2011) A review of sustainable cooling technologies in buildings. Renew Sustain Energy Rev 15(6):3112–3120
153. Bahadori MN (1985) An improved design of wind towers for natural ventilation and passive cooling. Sol Energy 35(2):119–129
154. Nejat P et al (2016) Evaluation of a two-sided windcatcher integrated with wing wall (as a new design) and comparison with a conventional windcatcher. Energy Build 126:287–300
155. Ding WT et al (2004) Smoke control based on a solar-assisted natural ventilation system. Build Environ 39(7):775–782
156. Cheng XD et al (2018) Study on optimizing design of solar chimney for natural ventilation and smoke exhaustion. Energy Build 170:145–156
157. Miao L, Chow CLJATE (2018) A study on window plume from a room fire to the cavity of a double-skin façade 129:230–241
158. Yang D et al (2013) A model for analysis of convection induced by stack effect in a shaft with warm airflow expelled from adjacent space. Energy Build 62:107–115
159. Shi L et al (2020) Solar chimney for a real building considering both energy-saving and fire safety—a case study. Energy Build 221:1–13

160. Haghighi AP, Maerefat M (2014) Solar ventilation and heating of buildings in sunny winter days using solar chimney. Sustain Cities Soc 10:72–79
161. Tan AYK, Wong NH (2013) Parameterization studies of solar chimneys in the tropics. Energies 6(1):145–163
162. Shi L et al (2018) Determining the influencing factors on the performance of solar chimney in buildings. Renew Sustain Energy Rev 88:223–238
163. Aldawoud AJE (2013) The influence of the atrium geometry on the building energy performance. Buildings 57:1–5
164. Ong K, Chow CJSE (2003) Performance of a solar chimney 74(1):1–17
165. Bernardes MDS, Voß A, Weinrebe GJSE (2003) Thermal and technical analyses of solar chimneys 75(6):511–524
166. Shi L et al (2019) Interaction effect of room opening and air inlet on solar chimney performance. Appl Therm Eng 159
167. Shi L (2018) Theoretical models for wall solar chimney under cooling and heating modes considering room configuration. Energy 165:925–938
168. Ascione F (2017) Energy conservation and renewable technologies for buildings to face the impact of the climate change and minimize the use of cooling. Sol Energy 154:34–100
169. Lomas KJJE (2007) Architectural design of an advanced naturally ventilated building form. Buildings 39(2):166–181
170. Linden P, Lane-Serff G, Smeed DJJOFM (1990) Emptying filling boxes: the fluid mechanics of natural ventilation 212:309–335
171. Gan GJTOC (2010) Interaction between wind and buoyancy effects in natural ventilation of buildings. Build Technol J 4(1)
172. Yang D, Guo YH (2016) Fluctuation of natural ventilation induced by nonlinear coupling between buoyancy and thermal mass. Int J Heat Mass Transf 96:218–230
173. Miyazaki T, Akisawa A, Nikai I (2011) The cooling performance of a building integrated evaporative cooling system driven by solar energy. Energy Build 43(9):2211–2218
174. Ghoulem M et al (2020) Design of a passive downdraught evaporative cooling windcatcher (PDEC-WC) system for greenhouses in hot climates. Energies 13(11):2934
175. Poshtiri AH, Mohabbati SM (2017) Performance analysis of wind catcher integrated with shower cooling system to meet thermal comfort conditions in buildings. J Clean Prod 148:452–466
176. Adekoya L (1992) Wind energy end-use: the performance characteristics of a rotating suction cowl. Renew Energy 2(4–5):385–389
177. Azhar S (2011) Building information modeling (BIM): trends, benefits, risks, and challenges for the AEC industry. Leadersh Manag Eng 11(3):241–252
178. Lu YJ et al (2017) Building information modeling (BIM) for green buildings: a critical review and future directions. Autom Constr 83:134–148
179. Utkucu D, Sözer H (2020) Interoperability and data exchange within BIM platform to evaluate building energy performance and indoor comfort. Autom Constr 116:103225
180. Afonso C, Oliveira A (2000) Solar chimneys: simulation and experiment. Energy Build 32(1):71–79

Chapter 2
Preliminaries

The critical factors of thermal and ventilation performance of natural ventilation systems, including airflow rate and temperature change, were concluded in Table 2.1. The cited literature comprises dissimilar investigation approaches to cope with different aspects. A comprehensive evaluation of the mentioned systems in the tables is complicated. It has been recognized that a stand-alone natural ventilation system can typically offer around 20–50% energy conservation, counting on the system design and external climatic conditions. For example, a solar-assisted passive system was recognized to lessen air-conditioning average electrical consumption approaching 10–20% [1, 2]. Ventilation rates achieved using the independent system usually cannot satisfy the overall ventilation requirements due to their dependence on the available ambient conditions [3–5]. Those solar-based natural ventilation systems highly rely on solar radiation, where cloudy weather conditions may hamper performance [6]. It is then essential for those solar-based systems to combine with systems not relying on solar radiation, such as wind-based systems [7]. That also emphasizes the significance of the combined natural ventilation systems in practical applications.

2.1 Research Methods

The studies in the ventilation areas are generally based on experimental methods coupled with numerical simulations [8–10]. Two fundamental models are usually adopted for detailed analytical and numerical analysis, including Bernoulli and Navier–Stokes equation. Chen et al. [11, 12] outlined those computer-aided prediction models, including empirical models, small-scale experimental models, full-scale experimental models, multi-zone network models, zonal models, and theoretical models (usually related to CFD and analytical models). It summarized the contributions of modern applications regarding the above simulation methods for practical design and research purposes.

L. Shi and H. Zhang, *Solar Chimney Applications in Buildings*, Green Energy and Technology, https://doi.org/10.1007/978-3-031-45218-5_2

Table 2.1 Natural ventilation applications

Performance / Natural ventilation techniques	Typical natural ventilation systems	Schematic diagram and airflow pattern	Indoor temperature adjustment (°C)	Range of relative humidity [%]	ACH	Volumetric/mass flow rate (l/s, m^3/s, m^3/h, or kg/s) [a] under given scenario	Energy-saving percentage (%)	Analysis method
1Thermal buoyancy-driven ventilation	Trombe wall	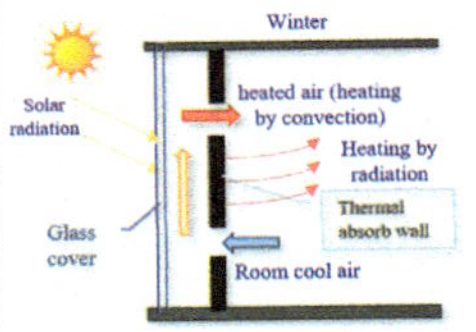	Drop 2–17 °C with evaporative cooling [36, 111]; increase around 1–2 °C [110]; reduce 9.1 °C [110]	28.3–28.8% [111]		0.01–0.02 kg/s [112]; 20–90 m^3/h [113]	30–80% [37, 93, 114–119]	Numerical study (CFD) and experimental study
	Double-skin façade (vent-skin walls)	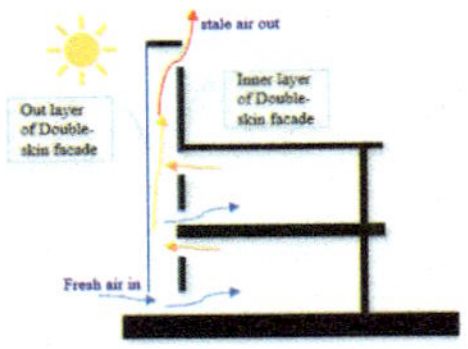	Rise 0.66–14.7 °C [59, 120–124]	50–95% [122, 125]	1.351 [120, 126, 127]; 4 [128]; 2–11 with solar chimney [129]	0.27–0.36 m^3/s with PV [37, 130]; 54 m^3/h with fan [42]; 0.042–0.255 kg/s [37, 122 –124, 131–133]; 16.1–22.4 l/s/m [81]	14%–75% (with shading device) [42, 59, 124, 134–136]	CFD [80, 137]; numerical study; experimental

(continued)

Table 2.1 (continued)

Performance / Natural ventilation techniques	Typical natural ventilation systems	Schematic diagram and airflow pattern	Indoor temperature adjustment (°C)	Range of relative humidity [%]	ACH	Volumetric/mass flow rate (l/s, m^3/s, m^3/h, or kg/s) [a] under given scenario	Energy-saving percentage (%)	Analysis method
	Solar (thermal) chimney/roof solar chimney	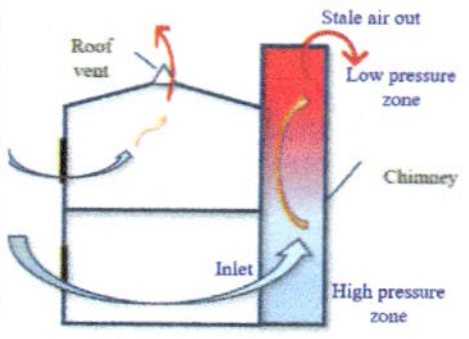	Drop 1.0–3.5 °C [3, 8, 20, 25, 28, 35, 37, 53, 121, 138–141]; 2.0–6.2 with water spaying roof [35]; 14 °C with water spraying; with EAHE drop 3.2–6.6 °C [142]; rise 7 °C [139]		2–6 [10, 27, 28, 143–147]; 8 [36] in coupling system of Trombe wall and water spraying system; 35–73 with wind tower [148]; 8–15 [145]	55–330 m^3/h [3, 8, 27, 28, 139, 140, 147–150]; 50–374 m^3/h [139]; 0.0085–0.5 kg/s [37, 151, 152]; roof solar chimney 0.08–0.15 $m^3/s/m^2$ [37, 153] or 10–100 m^3/h [150]; 0.014 kg/s [154]; 0.18–0.33 m^3/s[155]; 2.3 m^3/s [143]; 0.25–0.39 m/s [156]; double-air channel solar chimney 0.1072 kg/s [157]; 0.75–1.4 kg/s [148]	EHD: 620; HD: 101 [33]; 12–50% [1, 2, 59]	Numerical study; CFD; FDS; experimental investigation

(continued)

Table 2.1 (continued)

Performance / Natural ventilation techniques	Typical natural ventilation systems	Schematic diagram and airflow pattern	Indoor temperature adjustment (°C)	Range of relative humidity [%]	ACH	Volumetric/mass flow rate (l/s, m^3/s, m^3/h, or kg/s) [a] under given scenario	Energy-saving percentage (%)	Analysis method
	Solar walls (unglazed transpired solar façade and glazed transpired solar façade)	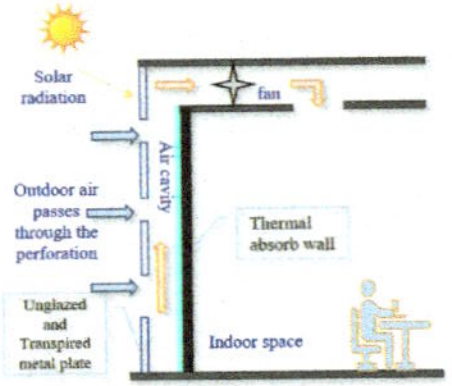	Unglazed transpired solar wall for space heating: rise 12–13 °C [37, 121, 158–160]		A linear relationship between ACH and solar intensity [10]	72–125 m^3/h/m^2 [160–162]	500–1000 kWh/m^2/year [160–162]	
	Atrium	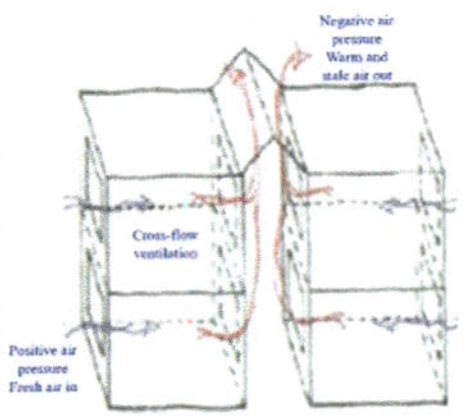	Rise up to 1.5–5 °C [88, 91, 163]; reduce 2.5–5 °C with solar PV [164]		A linear relationship between the ACH and wind speed/direction [52]	15000 l/s with a strong stack effect [32, 88]; 0.39–0.8 m^3/s [165; 7.8–73.7 l/s per m^2 [32]; 600–1000 l/s [95, 166]; 20–25 l/h [167]	40–80% [168]	

(continued)

Table 2.1 (continued)

Performance / Natural ventilation techniques	Typical natural ventilation systems	Schematic diagram and airflow pattern	Indoor temperature adjustment (°C)	Range of relative humidity [%]	ACH	Volumetric/mass flow rate (l/s, m^3/s, m^3/h, or kg/s) [ε] under given scenario	Energy-saving percentage (%)	Analysis method
Wind-induced ventilation	Wind tower	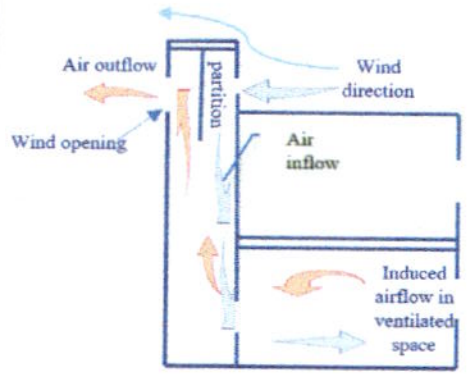	A drop of 4–25 °C [3, 33, 34, 48, 169–174]; reducing air temperature by 13% [175]; (with heat transfer device); reduce 5.2 °C with an SC and water spray [176]; reduce by up to 13.3 °C with evaporative cooling [177]; reduce 3.2 °C [178]	Increase of 19–22% [177, 179] with the use of wet columns; increase between 50 and 60% with evaporative cooling [170, 177]; decline by 15% [180]	1.75–73 [4, 49, 128, 181–184]; 9 with SC and water spray [176]; 3–20 with evaporative cooling for two-storey building [181]	0.75 g/s [185, 186]; 414 m^3/h [30]; 15.1 l/s/m^2 [5]; 0.3 m^3/s [172, 187]; 4–650 l/s [5, 21, 187–190, 3, 31, 34, 74, 92, 169]	[92] cooling energy reduction 43–61% [184] [33] EHD: 931; HD: 151; save 90% of the ventilation energy [176]	Field study [170, 180]; CFD [171, 191]; wind tunnel testing; experimental study
	Windcatcher	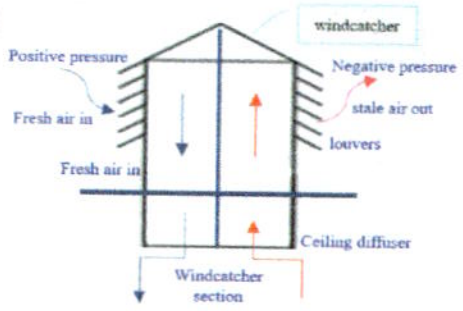	Drop 2–5 °C [34, 171, 174, 180, 192, 193] [194]; with cool sink drop 12 °C [195]	Decrease by 15% [180]; increase by 19% [171]	1.6–13.4 [196]	80–145 l/s [29, 31]		CFD; wind tunnel testing [174]; field research [196]; numerical study

(continued)

Table 2.1 (continued)

Performance / Natural ventilation techniques	Typical natural ventilation systems		Schematic diagram and airflow pattern	Indoor temperature adjustment (°C)	Range of relative humidity [%]	ACH	Volumetric/mass flow rate (l/s, m^3/s, m^3/h, or kg/s) [a] under given scenario	Energy-saving percentage (%)	Analysis method
	Fenestration	Single-sided ventilation				0.65–6.60 [197, 198]	0.054–0.550 m^3/s[198]		
		Cross-ventilation				5–22 [197, 199]		[200]	
	Wing walls		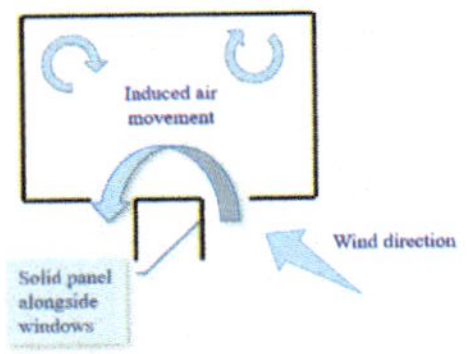			Induce fresh air supply up to 40% [31]; 0.54–5.9 [201]	The average indoor air velocity is 40% of the incident wind velocity, while without a wing wall, it is only 15% [202]	20% [203]	
	Wind cowls/scoops		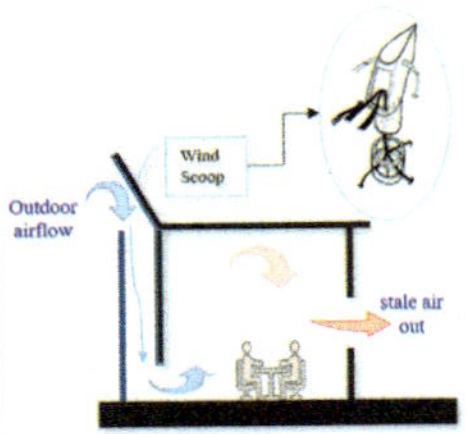				100–260 l/s [204, 205]		

(continued)

Table 2.1 (continued)

Performance Natural ventilation techniques	Typical natural ventilation systems	Schematic diagram and airflow pattern	Indoor temperature adjustment (°C)	Range of relative humidity [%]	ACH	Volumetric/mass flow rate (l/s, m^3/s, m^3/h, or kg/s) [a] under given scenario	Energy-saving percentage (%)	Analysis method
	Rotating wind cowl/turbine ventilators	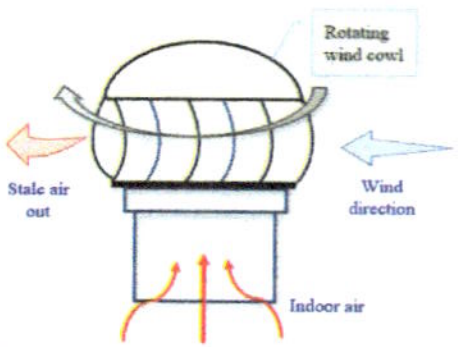				20–300 l/s [31, 206–211]; 270 m^3/h [212]; depencing on size and wind speed [213]		
	Exhaust cowls/roof vents/roof cowls	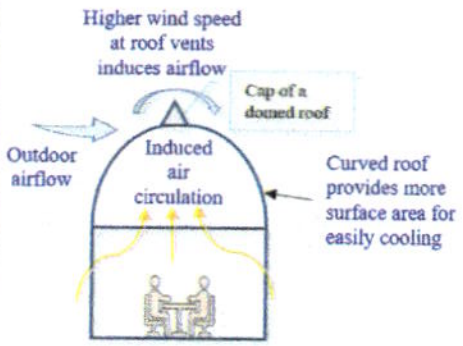			6 [214]	1764 m^3/h at wind speed of 2.5 m/s [31, 215]; <90 l/s [214]; various depending on size and wind speed [213]		Wind tunnel [214]

(continued)

Table 2.1 (continued)

Performance / Natural ventilation techniques	Typical natural ventilation systems	Schematic diagram and airflow pattern	Indoor temperature adjustment (°C)	Range of relative humidity [%]	ACH	Volumetric/mass flow rate (l/s, m^3/s, m^3/h, or kg/s) [a] under given scenario	Energy-saving percentage (%)	Analysis method
Heat modulation or amortization technique (modify heat gains)	Nocturnal cooling applications (night sky infrared radiation; day-time ventilation: shifting of day heat to night for removal)	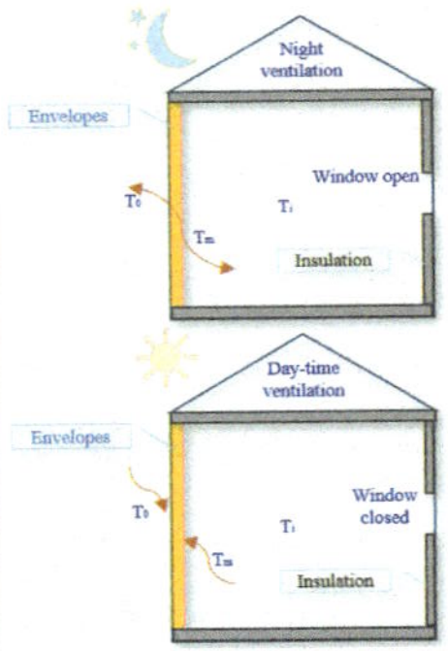	Reduced by up to 1.5–8 °C [34, 173, 200, 216–223]		10 [200, 220]		EHD: 1 918 [GW h]; HD: 312; HT/MT: 68 [33]; 11–90% [224, 225]	

Notes The unit of $m^3/s/m^2$ represents the volumetric airflow rate per square meter of the absorption wall. Nomenclature: solar wall: solar-assisted passive systems incorporating building-integrated photovoltaic, hot humid (HH), extreme hot humid (EHH), hot dry (HD), extreme hot dry (EHD), hot temperate (HT), and mild temperate (MT)

Computational fluid dynamic (CFD) modeling applications are suited for simulating natural convection and relevant temperature variations or velocity distributions. Unlike analytical models, the development of the prediction capability of CFD modeling in modeling ventilation performance can contribute to preferable appraisals to address the referred issues in developing optimization strategies [13]. Simultaneously, CFD simulation allows researchers to verify additional structure cases in a limited time for optimizing the performance of structures. Varela-Boydo et al. [14] used CFD modeling to establish 204 scale models in order to explore the configuration of the air outlet and air passage for the best performance of the windcatcher. Compared with a one-dimensional approach, the CFD technique can capture the prevailing context of flow phenomena like reverse flow, laminar, and turbulent, by a multi-dimensional analysis [15]. Naraghi and du Sordet [16] examined a correlation for calculating the volume flow rate for a solar roof chimney with the support of controlling parameters obtained using CFD. Nguyen and Wells [17] used CFD simulation to predict the performance of solar chimneys with different configurations through two indicators: flow rate and thermal efficiency. Using Fire Dynamics Simulator (FDS), Shi et al. [18, 19] developed an empirical model to predict the airflow rate through the air intakes under smoke exhaustion. It examined the impacts that differ with regard to the interaction mechanism between the air inlet and room windows on the solar chimney performance. Zamora and Kaiser [20] evaluated the effects of wind-buoyancy-induced airflow characteristics on a solar chimney's ventilation performance. Su et al. [21] evaluated the volume flow rate of a commercial wind tower device according to different external wind velocities using CFD, experiments, and far-field tests. The results critiqued that the buoyancy-driven flow shows an insignificant impact on the airflow rates of the wind tower at excessive outdoor wind speed. The details of methodology can be seen in Chap. 4.

2.2 Indicators of Performance

2.2.1 *Ventilation Rate*

Ventilation effectiveness evaluates the efficiency of the supplied fresh air commingled and distributed in the inhabited space [22]. The air exchange efficiency indicates the efficiency of the fresh air distributed in space. Table 2.1 gives information regarding the performance indicators from existing studies. Ventilation effectiveness indicates the efficiency of the stale air removed from the space [23]. The air exchange efficiency is measured by ACH or flow rate through the building. ACH is a measurement of the air volume gains or losses of a space divided by the volume of the space [24]. If the air supply volume to the inner spaces is adequate, the interior temperature could infinitely approach the ambient temperature. In general, natural ventilation can produce a considerable airflow rate economically on account of the use of natural phenomena without the aid of traditional energy resources. The ventilation rate is affected by many factors, including the intensity of wind or stack forces, the wind direction, and the resistance of the flow movement. The uncertainties of natural phenomena could result in the air change rate varying markedly. Moreover, the existence of uneven distribution of air movement to internal spaces leads to either inadequate ventilation or overventilation and excessive energy wastage.

The range of ACH values and the volumetric flow rate saw an apparent diversity in the same system in Table 2.1. This large fluctuation is due to different external conditions and research methods and the addition of different auxiliary systems to the same system. More specifically, ACH differs with regard to many design variables, such as structural geometry, the number of residents, the use of insulation, windows, and building function [8].

For those systems that equip with a vertical cavity, the higher the vertical height is, the higher the amount of induced air ventilation rates will be [25]. For example, intentional shafts can create more considerable differences between the top and bottom, increasing the airflow rate in a solar chimney, while the experimental results from Afonso and Oliveira [26] revealed that the chimney depth is more influential on the ventilation rate rather than its height. ACH will also increase with the intensified solar radiation and cavity depth [10, 27]. Although wind-driven can generate more considerable air movement inside the building than buoyancy-driven, a constant airflow can be maintained with appreciable buoyancy design. Khanal and Lei [8] summarized the research on a solar chimney in the past 20 years, especially for ventilation performance assessed by ACH and flow rate. It is argued that the previous studies were mainly divided into two research fields: one was to study the heat transfer process and fluid mechanics characteristics in the channel; the other was to study the flow and thermal behavior of the space inside the ventilated building.

On the other hand, wind-driven ventilation is peculiarly desirable in areas with relatively sufficient wind resources, and ventilation efficiency shows excellent diversity and heavily relies on local wind speed and direction. For example, an induced airflow rate of 140–330 m^3/h can be achieved at different ambient temperatures under solar radiation ranging from 200 to 1000 W/m^2 [28], while windcatcher supplies 106 and 145 l/s under direct wind with 0°and at wind speeds of 3 and 6 m/s, respectively. The air supplied was 106 l/s, and the air extracted was just over 80 l/s at a wind speed of 3 m/s [29]. Moreover, a wind tower could produce 414 m^3/h airflow rate, as shown in ref. [30]. Furthermore, the configuration of the system can affect the ventilation rate profoundly. Rotating chimney cowls produced 35–90 l/s depending on cowl size and wind speed [31].

2.2.2 Temperature

The majority of the research on natural ventilation methods was related to passive cooling [32]. It was known that the temperature reduction based on wind-driven seems to be better than that of buoyancy ventilation under certain environmental conditions. For example, in the studies taken by Oropeza-Perez and Ostergaard [33] and Jomehzadeh [34], temperature droppings induced by a wind tower up to 12–25 °C, while solar chimney can achieve a decrease of the internal temperature by roughly 1.0–3.5 °C [35]. The main reason is that the pressures generated by buoyancy are relatively low. Nonetheless, in terms of the space heating effect, the buoyancy-driven system performs even better. In stack ventilation, the indoor air temperature and the amount of ventilation are coupling relationships, which can be expressed by a quantitative coupling equation. The specific calculation methods are not listed and described in this article.

2.3 Assessment and Applicability

The initial intention of natural ventilation is to exhaust stagnant air, dilute indoor air pollutants, and regulate thermal comfort, heat dissipation, and air flow displacement. The challenges of natural ventilation are the difficulties of achieving a consistent airflow direction and the variation of the ventilation rate due to inconstant driving forces. They could result in the possibility of a low ACH during unfavorable climate conditions. Also, a well-sustained negative pressure is challenging to achieve the required natural ventilation performance. Excepting the interferences mentioned above, it is still a convincing alternative to reduce the operating frequency of conventional ventilation systems by using an appropriate design or integrating ventilation. Owing to these problems, taking advantage of a combination system that minimizes energy consumption becomes attractive.

The comparison of natural ventilation can be undertaken following the general concerns of the literature in Table 2.2. Through extensive literature, ten judgment criteria were identified for the pair-wise comparisons of the existing natural ventilation systems. However, there is no general criterion for evaluating the merits and availability of systems based on their performance. For example, heating systems capture and store the heat gain and retain the thermal within the building, while cooling systems provide aeration-cooling or protect the building from direct solar radiation and accelerate air circulation to remove warm air. However, not each natural ventilation system can have the dual function of simultaneously heating and cooling the space.

Additionally, the advantages and disadvantages of different systems under a specific scenario can also be observed in Table 2.1. Give a few examples, the Trombe wall is excellent in space cooling and heating convenience, especially since it can be combined with selected materials, various channel configurations, and different vents dimensions [6, 36]. However, the related reverse heat transfer needs to be resolved in some cases [37]. Notably, the performance of the DSF on acoustic insulation and thermal insulation is significant. However, fire safety has become an obstacle to the widespread use of DSF. DSF is usually used in conjunction with shading or sheltered devices to reduce direct solar gains into occupant spaces [38–42]. As for the solar chimney, temperature and air velocity are increased with higher solar radiation intensity. The air channel can be heated by solar energy to drive the air movement

Table 2.2 Overall comparison of typical natural ventilation technologies according to the ten judgment criteria

Major findings	Reference costs	Additional maintenance and operational costs	Simple structure	Space requirement	Applicability under various scenarios	Heating effect	Cooling effect	Human behavioral factors	Transparency—Architectural design	Fire safety
Trombe wall		√	√	√	√	√	√	×	×	×
Double-skin façade (vent-skin walls)	Standard 600 to 800 Euro/m^2; with openable exterior sashes 800 to 1300 Euro/m^2 [226]	√	√	√	√	√	√	√	√	×
Solar (thermal) chimney	501–1000 Euro [33]	√	√	√	√	√	√	×	×	×
Solar walls (unglazed transpired solar façade and glazed transpired solar façade)	NA	×	×	√	√	√	√	×	√	×
Atrium	NA	√	×	×	√	√	√	×	√	√
Wind tower	501–1000 Euro [33]	√	×	√	×	×	√	√	×	√

(continued)

Table 2.2 (continued)

Major findings	Reference costs	Additional maintenance and operational costs	Simple structure	Space requirement	Applicability under various scenarios	Heating effect	Cooling effect	Human behavioral factors	Transparency—Architectural design	Fire safety
Windcatcher	501–1000 Euro [33]	√	√	√	×	×	√	√	×	×
Single-sided ventilation (fenestration)	NA	√	√	√	×	×	√	√	√	√
Cross-ventilation (fenestration)	NA	√	√	√	×	×	√	√	√	√
Wing walls	NA	√	√	√	×	×	√	√	×	√
Wind cowls/ scoops	NA	√	√	√	×	×	√	√	×	×
Air inlets/wind vents/turbine ventilators	NA	√	√	√	×	×	√	√	×	×
Exhaust cowls/ roof vents/air inlets/roof cowls	NA	√	√	√	×	×	√	√	×	×
Nocturnal cooling (night sky infrared radiation)	0–30 Euro [33]	×	×	√	√	√	√	√	×	×
Day-time ventilation	0–30 Euro [33]	√	√	√	×	×	√	√	×	×

Note √ represents advantages, while × means the disadvantages or limitations of natural ventilation technology to take away the internal heat to achieve the cooling effect [43]. The problem that needs to be solved in the chimney effect is how to prevent flow reversal effects [15, 44]. Besides, it is generally considered that the solar chimney is not appropriate for regions with inadequate insolation or hot and humid climate since the air movement in SC is caused by intensifying temperature difference produced by direct or indirect solar gain.

In terms of wind-driven ventilation, different systems designs are based on the same principle [45–47]. Wind tower performs well under hot and dry conditions [34, 48–50]. Windcatcher performs exceptionally in achieving the desired IAQ and improving thermal comfort under a hot and humid climate [51]. Stagnant air is extracted from the intentional openings at the leeward side, owing to a negative pressure generated by the siphonic effect and stack effects [31]. A linear correlation was observed between wind speed and the extract flow rate. The suction effect of a wind-induced system depends on the prevailing wind speed and incident angle of the wind and the characteristics of cowls (regarding discharge coefficient) [52]. Thus, unstable and discontinuous wind and direction could limit ventilation performance. The dependent variables with the most straightforward influence on ventilation performance and thermal performance are elaborated on in the subsequent sections. These related causes affected different regions differently [53], and compromise may exist regarding thermal comfort level [54, 55].

2.3.1 Building Characteristics and Thermal Mass of Building Envelop

A close relationship has been found between the diversity of space geometry (i.e., building typology and position of building openings) and the air circulation system. Through energy simulation software (e.g., based on genetic algorithm, sequentially search technique particle swarm algorithm [56–58], and parametric study), researchers have conducted a series of case studies on approaching the impacts of structural characteristics on energy performance. The studies determined the optimal building features considering diverse parameters, such as facade type, opening area, building envelope, the height of different building parts, glazing type, building shape, the orientation of the windows, aspect ratio, and the windows-to-wall ratio [57, 59–72]. Pacheco et al. [73] stated that the physical configurations of ventilated spaces could affect the supply of adequate air, thermal conditioning, and moisture dissipation. Aflaki et al. [74] argued that reducing heat absorption from the environment is necessary to minimize energy loading. The study discussed the effect of architectural elements on ventilation performance in tropical climates, namely the building form, location, size of building openings, and features (i.e., orientation and layout).

Although numerous studies have been carried out on natural ventilation, their focus is on the single room to simplify it. There are relatively few studies on natural ventilation performance in multi-chambers. It is essential to approach practice by shifting the focus from those simple cases to practical applications, even though they have many influencing factors. The research on the effect of hot-pressed ventilation in multi-chamber space can be divided into two categories: the multi-chambers located at the same level. The other is the vertical distribution of the chambers, that is, multi-storey buildings. As for the first case, Lin et al. [75, 76] discussed how the interior geometry affected the ventilation flow and stated that the ventilation depends on the geometry and openings in the room and the size of adjacent rooms. The factors affecting the multi-floor space are relatively multitudinous. The multi-layer solar chimney with a vertical distribution model of three chambers was established by Punyasompun et al. [25] and validated by experimental data. Their study identified the optimal opening method for multi-layer solar chimneys. It verified that multi-layer buildings' indoor temperature using solar chimneys was 4–5 °C cooler than those without solar chimneys. Omrani et al. [77] summarized the evaluation methods and prediction tools of natural ventilation as a passive cooling strategy in the design process of multi-storey buildings.

Another consideration is the effects of diversity in configurations and dimensions of the passive system. Variables usually include the cavity width, temperature of the absorbing wall, internal air temperature, and chimney depth [16, 78]. Shi and Chew [79] provided a review on summarizing the influences of geometric factors and design variables on the solar chimney performance. Mathur et al. [27] assessed the effects of employing a solar chimney to enhance airflow rate and validated the proposed mathematical model with experimental outcomes. A trade-off between the absorber inclinations and stack height was suggested. Besides, Barbosa et al. [80] comprehensively reviewed the relevant literature on DSF and summarized various factors considered by DSF in practical applications and their impact on DSF performance. In addition to environmental factors, it is emphasized that the single-layer space cavity depth [81, 82], opening position, selection of internal and external skin materials, and the structural form of the cavity have a particularly significant impact on DSF performance. The conclusion shows that the narrow cavity is more conducive to the airflow generated by the buoyancy effect. The width of the cavity greater than one meter is not conducive to the chimney effect, and it also leads to an increase in the internal space heat gain.

Based on the balance between the ventilation rate to remove heat from the indoor space and the heat transfer from the cavity to the residential zone, Radhi et al. recommend a cavity depth between 0.7 and 1.2 m [81]. Gan [83] undertook the simulation on the buoyancy-driven ventilation. It was found that the prediction of ventilation rate and heat transfer coefficient were affected by space and cavity dimensions, inlet location, and heat distribution ratio [84]. Because the atrium is an ancient and distinctive form of architectural structure, its performance and building compositions are closely related. Aldawoud and Clark [85] compared the thermal behaviors of atria and courtyards with similar geometry under four types of local climates.

The coupling of the thermal mass of the ventilated building to buoyancy-driven ventilation has also successfully attracted the attention of many researchers and conducted a lot of research and development. Yang and Guo [86] established a theoretical model to analyze the nonlinearity of the coupling effect between the thermal mass of building envelopes and the buoyancy effect under an external heat source. The thermal profile and ventilation performance were discussed. Notably, it analyzed the uncoordinated ventilation rate fluctuations under the coupling effect. Similarly, such a nonlinear coupling relationship was also studied by Yam et al. [87]. Their theoretical model is divided into two cases. One is when the ventilation rate is constant; another is when the ventilation rate is a function of the temperature difference between indoors and outdoors. The control parameters for thermal behavior expression in these two cases are different.

2.3.2 *Location and Climate Conditions*

The ambient variables are essential in producing airflow. Some desired flow patterns depend on the available climatic factors. Tejero-González et al. [54] assessed the effectiveness of passive techniques with a comprehensive climate analysis. Bhamare et al. [53] provided a review of passive cooling techniques, including solar control techniques, heat modulation techniques, and heat dissipation techniques, considering the impacts of climate factors.

While naturally ventilated buildings may experience an ideal outdoor environment offering driven forces appropriately, the performances are distinctly fluctuating and dependent upon external variables. Consequently, the theoretical prediction of airflow patterns rarely occurs consistently along with time basis [54, 88], which does not mean it cannot offer the required performance. Ambient variables are potential constraints in natural ventilation applications, such as solar intensity, wind intensity, temperature, and humidity. For example, wind towers visible in ancient buildings in parts of Arabia can suck in the high-speed wind. The air travels down through the tower and drives heat away from building the interior by providing a constant air current and circulation. Contrarily, as the movable wind direction changes in a specific time range, the pressure field around the building also changes accordingly. In the hot season, energy consumption for refrigeration purposes is significant. Take an atrium, for example, Rojas [89] studied a five-storey atrium in the Mediterranean climate condition, showing that protecting solar radiance can give rise to a 75% decrease in cooling demand.

In comparison, night ventilation contributes to a reduction of 10%. However, the application of the atrium in the tropics could cause overheating problems, especially during periods of intense insolation. Excessive heat transfer from the broad surface of the glazing components at the upper space of the atrium could be occurred [90]. Mahyuddin et al. [91] reviewed the performance of natural ventilation using different atrium configurations and components in buildings under different climatic conditions. By changing the physical features and adopting an evaporative cooling system, a field investigation of a building with atrium space in the tropics was examined in a tropical environment in which five modes were developed [88, 92].

Besides, reducing energy consumption can also be realized by selecting different construction materials according to local conditions. Generally, insulation and thermal materials are employed in cold climates to reduce heat loss, whereas insulation and reflective materials are adopted to dissipate heat gain in hot climates [93].

2.3.3 Ambient Temperature

As demonstrated in the literature, changes in indoor and outdoor temperatures directly affect the efficiency of natural ventilation systems. For example, in hot and humid climates, the impacts of surrounding thermal conditions on the temperature and air distribution through atrium space are more significant than the heat loads inside the space [32, 94]. On the contrary, it is detrimental to employ the atrium system in scorching ambient temperatures (usually greater than 35 °C) [95]. If the temperature varies slightly between inside and outside, it can adversely affect the buoyancy-driven natural ventilation processes. Night ventilation contributes significantly to natural ventilation only when the exterior temperature drops obviously [23]. Nevertheless, the retardance of charging and discharging activities limits the performance and capacity of night ventilation on account of the reliance on the natural convection process and the dependency on structural thermal mass [96, 97].

2.3.4 External Wind

As one of the dominant driving forces, the external wind is critical in determining ventilation efficiency unless the building is fully sealed. Wind factors could also enhance the stack effect. Wang et al. [98] confirmed the influence of external wind on the optimal design parameters of wall-mounted solar chimneys. Since the additional wind driving force can avoid backflow in a larger cavity, the chimney gap should be increased by 0.2 m compared to the design size that only considers buoyancy driving. Simultaneously, the addition of wind reduces air resistance to enhance the solar chimney's performance. For example, wind force can drive the fresh air in an atrium space by enlarging the positive pressure employed at the windward side inlet toward the wind. It also weakens the negative pressure on the leeward outlet. The pressure at the air intake may vary from positive to negative pressure according to the fickle wind direction. When there is a reverse airflow at the roof outlet in the atrium, the inlet opening turns into an exhaust port. It results in an unexpected descending air movement and then produces a suppression of the stack effect caused by plum from the heating source [99].

Many studies have investigated the interaction between external wind and natural ventilation systems [100]. For example, Gratia et al. [101] analyzed the influences of wind on air temperature distribution inside the DSF cavity. It is carried out that the temperature inside the glazing cavity of DSF can be 50 °C higher than outside when there is no wind influence; the temperature in the cavity is 30 °C higher than outside at a wind speed of 4 m/s. Gratia and De Herde [102] further studied the airflow pattern under the external wind. The results showed that when the incident wind is perpendicular to the skin of the DSF, the airflows between the different floors are similar. In this case, the wind effect on the upper floors was more significant than the buoyancy effect. In contrast, when the wind was blowing from the opposite direction, the air was extracted into the cavity on the lower floors. Saadataian et al. [103] emphasized the mechanisms of wind towers under the buoyancy effect and wind-driven flow. It argued that wind-driven flow is an essential variable in determining the ventilation rate when the wind source is adequate. Also, the study examined the attributes and configurations that can achieve optimal performance. Hughes and Ghani [5] presented a CFD simulation of a commercial wind tower. It explored the effects of the variations of external wind on the capability of passive ventilation rate. Similarly, Jones and Kirby [104] also applied a semi-empirical model to determine the effects of the varying external wind features on the performance of the wind tower.

2.3.5 Solar Radiation

Solar radiance is necessary to demand space heating during the cold weather and the encouragement of buoyancy ventilation during the hot season. Menchaca-Brandan et al. [105] affirmed the importance of accounting for radiative effects on airflow modeling, especially regarding the air temperature profile. It is pointed out that the ignorance of radiation results in the air temperature of occupant space being lower by 2–4 °C than the cases with the consideration. The air velocity was higher in simulations considering the radiation [106]. Hence, neglecting radiative heat transfer could bring an unreliable prediction of thermal comfort level. In other words, theoretical models or numerical simulations that omit radiation may mislead room temperature distribution predictions. According to the research on an atrium, the solar radiance that penetrates through the glazing walls or roof has the primary role in producing thermal stratification in atrium space [107]. When the atrium/courtyard is exposed to extreme solar radiation for a long time, it places a severe problem on the energy conversation [108]. The overheated air was conducted into the adjacent spaces, which led to extra energy consumption to avert thermal discomfort.

Furthermore, the concurrent influences of solar radiance and external wind on the same natural ventilation system were also compared by some studies. For example, Shi [7] investigated the interaction of a solar chimney with solar radiation and external wind under different wind angles (the angle between the incident wind and the outward normal of the wall with the window) numerically and theoretically. A windward situation ($0° < \alpha < 90°$) was suggested, but it was surprising to know that the leeward scenario with $\alpha = 180°$ presents a narrow effect on the performance of the solar chimney. The study affirmed that solar chimney performance is utterly dependent on the external wind rather than solar radiation and developed a mathematical model to predict the critical wind velocity.

2.3.6 *Relative Humidity*

There are not many direct studies addressing relative humidity's influences on natural ventilation performance. Humidity and condensation prevention are two considerations related to thermal comfort. Insufficient ventilation rate may result in more significant indoor relative humidity and moisture accumulation. Simultaneously, the significant increment in relative humidity will lead to a diminution in airflow rate at the chimney inlet and outlet [109]. Besides, the high relative humidity could also lower the efficacy of a passive cooling system [54]. The quantitative influence and how the relative humidity affects the natural ventilation systems can be critical for future work.

References

1. Khedari J, Rachapradit N, Hirunlabh J (2003) Field study of performance of solar chimney with air-conditioned building. Energy 28(11):1099–1114
2. Miyazaki T, Akisawa A, Kashiwagi TJRE (2006) The effects of solar chimneys on thermal load mitigation of office buildings under the Japanese climate 31(7):987–1010
3. Hughes BR, Calautit JK, Ghani SA (2012) The development of commercial wind towers for natural ventilation: a review. Appl Energy 92:606–627
4. Elmualim AA, Awbi HB (2002) Wind tunnel and CFD investigation of the performance of windcatcher ventilation systems 1(1):53–64
5. Hughes BR, Ghani SAJE (2008) Investigation of a windvent passive ventilation device against current fresh air supply recommendations. Buildings 40(9):1651–1659
6. Zhang LL et al (2020) Trombe wall for a residential building in Sichuan-Tibet alpine valley—a case study. Renew Energy 156:31–46
7. Shi L (2019) Impacts of wind on solar chimney performance in a building. Energy 185:55–67
8. Khanal R, Lei C (2011) Solar chimney—a passive strategy for natural ventilation. Energy Build 43(8):1811–1819
9. Zhai ZJ et al (2007) Evaluation of various turbulence models in predicting airflow and turbulence in enclosed environments by CFD: part 1—summary of prevalent turbulence models 13(6):853–870
10. Park D, Battaglia F (2015) Application of a wall-solar chimney for passive ventilation of dwellings. J Solar Energy Eng-Trans ASME 137(6)
11. Chen QY (2009) Ventilation performance prediction for buildings: a method overview and recent applications. Build Environ 44(4):848–858
12. Chen Q et al (2010) Ventilation performance prediction for buildings: model assessment 45(2):295–303
13. Jiru TE, Bitsuamlak GT (2010) Application of CFD in modelling wind-induced natural ventilation of buildings—a review. Int J Vent 9(2):131–147
14. Varela-Boydo C, Moya S, Watkins R (2020) Analysis of traditional windcatchers and the effects produced by changing the size, shape, and position of the outlet opening. J Build Eng 33:101828
15. Khanal R, Lei CW (2012) Flow reversal effects on buoyancy induced air flow in a solar chimney. Sol Energy 86(9):2783–2794
16. Naraghi MH, du Sordet RC (2011) Stack effect correlation for rooftop solar chimneys. In: ASME 2011 international mechanical engineering congress and exposition. American Society of Mechanical Engineers Digital Collection

17. Nguyen Y, Wells J (2020) A numerical study on induced flowrate and thermal efficiency of a solar chimney with horizontal absorber surface for ventilation of buildings. J Build Eng 28:101050
18. Shi L et al (2019) Interaction effect of room opening and air inlet on solar chimney performance. Appl Thermal Eng 159
19. Shi L, Zhang GM (2016) An empirical model to predict the performance of typical solar chimneys considering both room and cavity configurations. Build Environ 103:250–261
20. Zamora B, Kaiser AS (2010) Numerical study on mixed buoyancy-wind driving induced flow in a solar chimney for building ventilation. Renew Energy 35(9):2080–2088
21. Su Y et al (2008) Experimental and CFD study of ventilation flow rate of a monodraught™ windcatcher 40(6):1110–1116
22. Zhai XQ, Song ZP, Wang RZ (2011) A review for the applications of solar chimneys in buildings. Renew Sustain Energy Rev 15(8):3757–3767
23. Hughes BR, Chaudhry HN, Ghani SA (2011) A review of sustainable cooling technologies in buildings. Renew Sustain Energy Rev 15(6):3112–3120
24. Ashrae A (2016) Ventilation and acceptable indoor air quality in low-rise residential buildings
25. Punyasompun S et al (2009) Investigation on the application of solar chimney for multi-storey buildings. Renew Energy 34(12):2545–2561
26. Afonso C, Oliveira AJE (2000) Solar chimneys: simulation and experiment. Buildings 32(1):71–79
27. Mathur J et al (2006) Experimental investigations on solar chimney for room ventilation 80(8):927–935
28. Bansal N et al (1993) Solar chimney for enhanced stack ventilation. 28(3):373–377
29. Li L, Mak CJB (2007) The assessment of the performance of a windcatcher system using computational fluid dynamics. Environment 42(3):1135–1141
30. Abdallah ASH et al (2014) Parametric investigation of solar chimney with new cooling tower integrated in a single room for New Assiut city. Egypt Clim 5(2–3):92
31. Khan N, Su Y, Riffat SB (2008) A review on wind driven ventilation techniques. Energy Build 40(8):1586–1604
32. Mouriki E (2009) Solar-assisted hybrid ventilation in an institutional building. Concordia University
33. Oropeza-Perez I, Ostergaard PA (2018) Active and passive cooling methods for dwellings: a review. Renew Sustain Energy Rev 82:531–544
34. Jomehzadeh F et al (2017) A review on windcatcher for passive cooling and natural ventilation in buildings, part 1: indoor air quality and thermal comfort assessment. Renew Sustain Energy Rev 70:736–756
35. Chungloo S, Limmeechokchai B (2007) Application of passive cooling systems in the hot and humid climate: the case study of solar chimney and wetted roof in Thailand. Build Environ 42(9):3341–3351
36. Rabani M et al (2015) Empirical investigation of the cooling performance of a new designed Trombe wall in combination with solar chimney and water spraying system. Energy Build 102:45–57
37. Chan H-Y, Riffat SB, Zhu J (2010) Review of passive solar heating and cooling technologies. Renew Sustain Energy Rev 14(2):781–789
38. Tao Y et al (2020) Ventilation performance of a naturally ventilated double-skin façadein buildings. Renew Energy 167:184–198
39. Haase M, da Silva FM, Amato A (2009) Simulation of ventilated facades in hot and humid climates. Energy Build 41(4):361–373
40. Chou S et al (2009) A study on the effects of double skin façades on the energy management in buildings. 50(9):2275–2281
41. Gratia E, De Herde AJE (2007) The most efficient position of shading devices in a double-skin facade. Buildings 39(3):364–373
42. Jiru TE et al (2011) Airflow and heat transfer in double skin facades. 43(10):2760–2766

43. Haghighi AP, Maerefat M (2014) Solar ventilation and heating of buildings in sunny winter days using solar chimney. Sustain Cities Soc 10:72–79
44. Ren XH et al (2019) Thermal driven natural convective flows inside the solar chimney flush-mounted with discrete heating sources: reversal and cooperative flow dynamics. Renew Energy 138:354–367
45. Bahadori MN (1985) An improved design of wind towers for natural ventilation and passive cooling. Sol Energy 35(2):119–129
46. Bahadori MN, Pakzad AR (2002) Performance evaluation of new designs of wind towers. In: ASME 2002 joint US-European fluids engineering division conference. American Society of Mechanical Engineers Digital Collection
47. Bahadori M, Mazidi M, Dehghani AJRE (2008) Experimental investigation of new designs of wind towers 33(10):2273–2281
48. Kalantar V (2009) Numerical simulation of cooling performance of wind tower (Baud-Geer) in hot and arid region. Renew Energy 34(1):246–254
49. Bahadori MNJRE (1994) Viability of wind towers in achieving summer comfort in the hot arid regions of the Middle East 5(5–8):879–892
50. Bouchahm Y, Bourbia F, Belhamri A (2011) Performance analysis and improvement of the use of wind tower in hot dry climate. Renew Energy 36(3):898–906
51. Haw LC et al (2012) Empirical study of a wind-induced natural ventilation tower under hot and humid climatic conditions. Energy Build 52:28–38
52. Horan JM, Finn DPJE (2008) Sensitivity of air change rates in a naturally ventilated atrium space subject to variations in external wind speed and direction. Buildings 40(8):1577–1585
53. Bhamare DK, Rathod MK, Banerjee J (2019) Passive cooling techniques for building and their applicability in different climatic zones-the state of art. Energy Build 198:467–490
54. Tejero-Gonzalez A et al (2016) Assessing the applicability of passive cooling and heating techniques through climate factors: an overview. Renew Sustain Energy Rev 65:727–742
55. Chiesa G, Huberman N, Pearlmutter D (2019) Geo-climatic potential of direct evaporative cooling in the Mediterranean region: a comparison of key performance indicators. Build Environ 151:318–337
56. Tuhus-Dubrow D, Krarti MJAT (2009) Comparative analysis of optimization approaches to design building envelope for residential buildings 115(2):554–563
57. Tuhus-Dubrow D, Krarti MJB (2010) Genetic-algorithm based approach to optimize building envelope design for residential buildings. Environment 45(7):1574–1581
58. Bichiou Y, Krarti MJE (2011) Optimization of envelope and HVAC systems selection for residential buildings. Buildings 43(12):3373–3382
59. Sadineni SB, Madala S, Boehm RF (2011) Passive building energy savings: a review of building envelope components. Renew Sustain Energy Rev 15(8):3617–3631
60. Yi YK, Malkawi AMJAIC (2009) Optimizing building form for energy performance based on hierarchical geometry relation. 18(6):825–833
61. Capeluto IGJE (2003) Energy performance of the self-shading building envelope. Buildings 35(3):327–336
62. Adamski MJB (2007) Optimization of the form of a building on an oval base. Environment 42(4):1632–1643
63. Kämpf JH, Robinson DJE (2010) Optimisation of building form for solar energy utilisation using constrained evolutionary algorithms. Buildings 42(6):807–814
64. Kämpf JH et al (2010) Optimisation of buildings solar irradiation availability 84(4):596–603
65. Hachem C, Athienitis A, Fazio PJSE (2011) Parametric investigation of geometric form effects on solar potential of housing units 85(9):1864–1877
66. Hachem C et al (2012) Evaluation of energy supply and demand in solar neighborhood 49:335–347
67. Depecker P et al (2001) Design of buildings shape and energetic consumption 36(5):627–635
68. Ouarghi R, Krarti MJAT (2006) Building shape optimization using neural network and genetic algorithm approach 112(1)

69. Wang W, Rivard H, Zmeureanu R (2006) Floor shape optimization for green building design. Adv Eng Inform 20(4):363–378
70. Caldas LG, Norford LKJJSEE (2003) Genetic algorithms for optimization of building envelopes and the design and control of HVAC systems 125(3):343–351
71. Wang W, Rivard H, Zmeureanu RJAEI (2005) An object-oriented framework for simulation-based green building design optimization with genetic algorithms 19(1):5–23
72. Wang Y, Wei C. Design optimization of office building envelope based on quantum genetic algorithm for energy conservation. J Build Eng 35:102048
73. Pacheco R, Ordonez J, Martinez G (2012) Energy efficient design of building: a review. Renew Sustain Energy Rev 16(6):3559–3573
74. Aflaki A et al (2015) A review on natural ventilation applications through building facade components and ventilation openings in tropical climates. Energy Build 101:153–162
75. Lin YJP, Linden PF (2002) Buoyancy-driven ventilation between two chambers. J Fluid Mech 463:293–312
76. Lin YJP, Tsai TH (2013) The push-type displacement ventilation in two series-connected chambers. Build Environ 62:89–101
77. Omrani S et al (2017) Natural ventilation in multi-storey buildings: design process and review of evaluation tools. Build Environ 116:182–194
78. Tan AYK, Wong NH (2013) Parameterization studies of solar chimneys in the tropics. Energies 6(1):145–163
79. Shi L, Chew MYL (2012) A review on sustainable design of renewable energy systems. Renew Sustain Energy Rev 16(1):192–207
80. Barbosa S, Ip K (2014) Perspectives of double skin facades for naturally ventilated buildings: a review. Renew Sustain Energy Rev 40:1019–1029
81. Radhi H, Sharples S, Fikiry F (2013) Will multi-facade systems reduce cooling energy in fully glazed buildings? A scoping study of UAE buildings. Energy Build 56:179–188
82. Rahmani B, Kandar MZ, Rahmani PJWASJ (2012) How double skin façade's air-gap sizes effect on lowering solar heat gain in tropical climate 18(6):774–778
83. Gan GJE (2010) Simulation of buoyancy-driven natural ventilation of buildings—impact of computational domain. Buildings 42(8):1290–1300
84. Liu TQ, Lee WL (2019) Influence of window opening degree on natural ventilation performance of residential buildings in Hong Kong. Sci Technol Built Environ
85. Aldawoud A, Clark RJE (2008) Comparative analysis of energy performance between courtyard and atrium in buildings. Buildings 40(3):209–214
86. Yang D, Guo YH (2016) Fluctuation of natural ventilation induced by nonlinear coupling between buoyancy and thermal mass. Int J Heat Mass Transf 96:218–230
87. Yam J, Li YG, Zheng ZH (2003) Nonlinear coupling between thermal mass and natural ventilation in buildings. Int J Heat Mass Transf 46(7):1251–1264
88. Moosavi L et al (2014) Thermal performance of atria: an overview of natural ventilation effective designs. Renew Sustain Energy Rev 34:654–670
89. Palma Rojas DJIJOL-CT (2014) Atrium building design: key aspects to improve their thermal performance on the Mediterranean climate of Santiago de Chile 9(4):327–330
90. Thirugnanasambandam M, Iniyan S, Goic R (2010) A review of solar thermal technologies☆. Renew Sustain Energy Rev 14(1):312–322
91. Moosavi L, Ghafar N, Mahyuddin N (2016) Investigation of thermal performance for atria: a method overview. 4th International building control conference 2016 (IBCC 2016), p 66
92. Moosavi L et al (2018) Numerical prediction of thermal and airflow conditions of a naturally ventilated atrium and validation of CFD models. J Renew Sustain Energy 10(6)
93. Stazi F, Mastrucci A, di Perna C (2012) The behaviour of solar walls in residential buildings with different insulation levels: an experimental and numerical study. Energy Build 47:217–229
94. Liu PC, Lin HT, Chou JH (2009) Evaluation of buoyancy-driven ventilation in atrium buildings using computational fluid dynamics and reduced-scale air model. Build Environ 44(9):1970–1979

95. Wang X, Huang C, Cao WW (2009) Mathematical modeling and experimental study on vertical temperature distribution of hybrid ventilation in an atrium building. Energy Build 41(9):907–914
96. Oropeza-Perez I, Østergaard PAJAE (2014) Energy saving potential of utilizing natural ventilation under warm conditions—a case study of Mexico 130:20–32
97. Chaudhry HN, Calautit JK, Hughes B (2015) Computational analysis of a wind tower assisted passive cooling technology for the built environment. J Build Eng 1:63–71
98. Wang Q et al (2020) External wind on the optimum designing parameters of a wall solar chimney in building. Sustain Energy Technol Assess 42:100842
99. Ray SD et al (2014) Experimental characterization of full-scale naturally ventilated atrium and validation of CFD simulations. Energy Build 69:285–291
100. Chartier Y, Pessoa-Silva C (2009) Natural ventilation for infection control in health-care settings. World Health Organization
101. Gratia E, De Herde AJE (2007) Greenhouse effect in double-skin facade. Buildings 39(2):199–211
102. Gratia E, De Herde A (2004) Is day natural ventilation still possible in office buildings with a double-skin facade? Build Environ 39(4):399–409
103. Saadatian O et al (2012) Review of windcatcher technologies. Renew Sustain Energy Rev 16(3):1477–1495
104. Jones BM, Kirby RJB (2009) Quantifying the performance of a top–down natural ventilation windcatcher™. Environment 44(9):1925–1934
105. Menchaca-Brandan MA, Espinosa FAD, Glicksman LR (2017) The influence of radiation heat transfer on the prediction of air flows in rooms under natural ventilation. Energy Build 138:530–538
106. Du L, Ping L, Yongming C (2020) Study and analysis of air flow characteristics in Trombe wall. Renew Energy 162:234–241
107. Li R, Pitts A (2006) Thermal comfort and environmental modelling in atrium buildings. In: Proceedings PLEA2006—the 23rd conference on passive and low energy architecture. Geneva, Switzerland
108. Pan Y et al (2010) Study on simulation methods of atrium building cooling load in hot and humid regions. 42(10):1654–1660
109. Sudprasert S, Chinsorranant C, Rattanadecho P (2016) Numerical study of vertical solar chimneys with moist air in a hot and humid climate. Int J Heat Mass Transf 102:645–656
110. Hernández-López I et al (2020) Thermal performance of a solar façade system for building ventilation in the southeast of Mexico. Renew Energy 145:294–307
111. Rabani M, Kalantar V, Rabani M (2019) Passive cooling performance of a test room equipped with normal and new designed Trombe walls: a numerical approach. Sustain Energy Technol Assess 33:69–82
112. Hirunlabh J et al (1999) Study of natural ventilation of houses by a metallic solar wall under tropical climate. Renew Energy 18(1):109–119
113. Khedari J et al (1998) The modified Trombe wall: a simple ventilation means and an efficient insulating material 19(2):104–110
114. Saadatian O et al (2012) Trombe walls: a review of opportunities and challenges in research and development. Renew Sustain Energy Rev 16(8):6340–6351
115. Burek SAM, Habeb A (2007) Air flow and thermal efficiency characteristics in solar chimneys and Trombe walls. Energy Build 39(2):128–135
116. Modirrousta S, Boostani H (2016) Analysis of atrium pattern, trombe wall and solar greenhouse on energy efficiency. ICSDEC 2016—integrating data science, construction and sustainability, vol 145, pp 1549–1556
117. Bie Y et al (2012) Theoretical study on thermal performance of Trombe wall combined with solar air collector. Renew Sustain Energy Ii, Pts 1–4:512–515,208-+
118. Bevilacqua P et al (2019) Are Trombe walls suitable passive systems for the reduction of the yearly building energy requirements? Energy 185:554–566

119. Hu ZT et al (2017) A review on the application of Trombe wall system in buildings. Renew Sustain Energy Rev 70:976–987
120. Amirifard F, Sharif SA, Nasiri F (2019) Application of passive measures for energy conservation in buildings—a review. Adv Build Energy Res 13(2):282–315
121. Nanda AK, Panigrahi CK (2016) A state-of-the-art review of solar passive building system for heating or cooling purpose. Front Energy 10(3):347–354
122. Prieto A et al (2017) 25 Years of cooling research in office buildings: review for the integration of cooling strategies into the building facade (1990–2014). Renew Sustain Energy Rev 71:89–102
123. Zhang TT, Yang HX (2019) Flow and heat transfer characteristics of natural convection in vertical air channels of double-skin solar facades. Appl Energy 242:107–120
124. Sanchez E et al (2016) Influence of natural ventilation due to buoyancy and heat transfer in the energy efficiency of a double skin facade building. Energy Sustain Dev 33:139–148
125. Hien WN et al (2005) Effects of double glazed facade on energy consumption, thermal comfort and condensation for a typical office building in Singapore 37(6):563–572
126. Brandl D et al (2014) Analysis of ventilation effects and the thermal behaviour of multifunctional facade elements with 3D CFD models. Energy Build 85:305–320
127. Hensen J, Bartak M, Drkal FJAT (2002) Modeling and simulation of a double-skin facade system 108(2):1251–1259
128. Gratia E, De Herde A (2007) Guidelines for improving natural daytime ventilation in an office building with a double-skin facade. Sol Energy 81(4):435–448
129. Ding W et al (2005) Natural ventilation performance of a double-skin façade with a solar chimney 37(4):411–418
130. Gan GH (2006) Simulation of buoyancy-induced flow in open cavities for natural ventilation. Energy Build 38(5):410–420
131. Pak R, Ocak Z, Sorguven E (2018) Developing a passive house with a double-skin envelope based on energy and airflow performance. Build Simul 11(2):373–388
132. Gagliano A, Nocera F, Aneli S (2016) Thermodynamic analysis of ventilated facades under different wind conditions in summer period. Energy Build 122:131–139
133. Wang Y, Chen Y, Li C (2019) Airflow modeling based on zonal method for natural ventilated double skin façade with Venetian blinds. Energy Build 191:211–223
134. Saroglou T et al (2019) A study of different envelope scenarios towards low carbon high-rise buildings in the Mediterranean climate—can DSF be part of the solution? Renew Sustain Energy Rev 113
135. Ghaffarianhoseini A et al (2016) Exploring the advantages and challenges of double-skin facades (DSFs). Renew Sustain Energy Rev 60:1052–1065
136. Chan A et al (2009) Investigation on energy performance of double skin façade in Hong Kong 41(11):1135–1142
137. De Gracia A et al (2013) Numerical modelling of ventilated facades: a review. Renew Sustain Energy Rev 22:539–549
138. Monghasemi N, Vadiee A (2018) A review of solar chimney integrated systems for space heating and cooling application. Renew Sustain Energy Rev 81:2714–2730
139. Arce J et al (2009) Experimental study for natural ventilation on a solar chimney. Renew Energy 34(12):2928–2934
140. Omrany H et al (2016) Application of passive wall systems for improving the energy efficiency in buildings: a comprehensive review. Renew Sustain Energy Rev 62:1252–1269
141. Rabani R et al (2014) Numerical simulation of an innovated building cooling system with combination of solar chimney and water spraying system. Heat Mass Transf 50(11):1609–1625
142. Ramírez-Dávila L et al (2014) Numerical study of earth-to-air heat exchanger for three different climates 76:238–248
143. AboulNaga M, Abdrabboh SJRE (2000) Improving night ventilation into low-rise buildings in hot-arid climates exploring a combined wall–roof solar chimney 19(1–2):47–54

144. Hosien MA, Selim SM (2017) Effects of the geometrical and operational parameters and alternative outer cover materials on the performance of solar chimney used for natural ventilation. Energy Build 138:355–367
145. Khedari J et al (2000) Ventilation impact of a solar chimney on indoor temperature fluctuation and air change in a school building 32(1):89–93
146. Imran AA, Jalil JM, Ahmed ST (2015) Induced flow for ventilation and cooling by a solar chimney. Renew Energy 78:236–244
147. Zavala-Guillén I et al (2019) Ventilation potential of an absorber-partitioned air channel solar chimney for diurnal use under Mexican climate conditions. Appl Therm Eng 149:807–821
148. Bansal NK, Mathur R, Bhandari MS (1994) A study of solar chimney assisted wind tower system for natural ventilation in buildings. Build Environ 29(4):495–500
149. Jiménez-Xamán C et al (2019) Solar chimneys with a phase change material for buildings: an overview using CFD and global energy balance. Energy Build 186:384–404
150. Khedari J et al (2000) Field measurements of performance of roof solar collector. Energy Build 31(3):171–178
151. Xaman J et al (2019) Transient thermal analysis of a solar chimney for buildings with three different types of absorbing materials: copper plate/PCM/concrete wall. Renew Energy 136:139–158
152. Martı J, Heras-Celemin MJSE (2007) Dynamic physical model for a solar chimney 81(5):614–622
153. Khedari J, Hirunlabh J, Bunnag TJRE (1996) Expertmental study of a roof solar collector towards the natural ventilation of new habitations 8(1–4):335–338
154. Ong KS (2003) A mathematical model of a solar chimney. Renew Energy 28(7):1047–1060
155. Naraghi MH, Blanchard S (2015) Twenty-four hour simulation of solar chimneys. Energy Build 94:218–226
156. Ong K, Chow CJSE (2003) Performance of a solar chimney 74(1):1–17
157. Zavala-Guillen I et al (2018) Numerical study of the optimum width of 2a diurnal double air channel solar chimney. Energy 147:403–417
158. Saadatian O et al (2013) A state of the art review of solar walls: concepts and applications. J Build Phys 37(1):55–79
159. Cekon M, Slavik R (2017) A non-ventilated solar facade concept based on selective and transparent insulation material integration: an experimental study. Energies 10(6)
160. Hollick JCJRE (1994) Unglazed solar wall air heaters 5(1–4):415–421
161. Cali A et al (1999) A report of task 14 air systems working group: low cost high performance solar air heating systems using perforated absorbers
162. Hollick JJRE (1998) Solar cogeneration panels 15(1–4):195–200
163. Ab Ghafar N, Gadi M, Adam M (2019) Evaluation of thermal and solar performance in atrium buildings using sequential simulation. Energy Environ 30(6):969–990
164. Gan GH, Riffat SB (2004) CFD modelling of air flow and thermal performance of an atrium integrated with photovoltaics. Build Environ 39(7):735–748
165. Wang F, Abdullah AHJIJOL-CT (2011) Investigating thermal conditions in a tropic atrium employing CFD and DTM techniques 6(3):171–186
166. Mouriki E et al (2008) Full-scale study of an atrium integrated with a hybrid ventilation system. In: Proceedings of the 3rd Canadian solar buildings conference. Fredericton, Canada
167. Tanaka H et al (2009) Thermal characteristics of a double-glazed external wall system with roll screen in cooling season. Build Environ 44(7):1509–1516
168. Krarti M (2014) Estimation of lighting energy savings from atrium daylighting for office buildings. In: Proceedings of the ASME 8th international conference on energy sustainability, 2014, vol 2
169. Calautit JK, Hughes BR (2014) Integration and application of passive cooling within a wind tower for hot climates. HVAC&R Res 20(7):722–730
170. Elmualim AAJIJOV (2006) Verification of design calculations of a wind catcher/tower natural ventilation system with performance testing in a real building 4(4):393–404

171. Zarandi MMJET (2009) Analysis on Iranian wind catcher and its effect on natural ventilation as a solution towards sustainable architecture (Case Study: Yazd) 54:574–579
172. Badran AAJE (2003) Performance of cool towers under various climates in Jordan. Buildings 35(10):1031–1035
173. Gupta N, Tiwari GN (2016) Review of passive heating/cooling systems of buildings. Energy Sci Eng 4(5):305–333
174. Elmualim AA (2006) Effect of damper and heat source on wind catcher natural ventilation performance. Energy Build 38(8):939–948
175. Calautit JK, Hughes BR, Ghani SAJASR (2013) A numerical investigation into the feasibility of integrating green building technologies into row houses in the Middle East 56(4):279–296
176. Moosavi L et al (2020) New design for solar chimney with integrated windcatcher for space cooling and ventilation. Build Environ 106785
177. Ghoulem M et al (2020) Design of a passive downdraught evaporative cooling windcatcher (PDEC-WC) system for greenhouses in hot climates. Energies 13(11):2934
178. Sadeghi M et al (2020) Comfort cooling by wind towers in the Australian residential context–experimental wind tunnel study of comfort. J Wind Eng Ind Aerodyn 196:104014
179. Saffari H, Hosseinnia SJE (2009) Two-phase Euler-Lagrange CFD simulation of evaporative cooling in a wind tower. Buildings 41(9):991–1000
180. Priya RS, Sundarraja M, Radhakrishnan SJEE (2012) Experimental study on the thermal performance of a traditional house with one-sided wind catcher during summer and winter 5(4):483–496
181. Poshtiri AH, Mohabbati SM (2017) Performance analysis of wind catcher integrated with shower cooling system to meet thermal comfort conditions in buildings. J Clean Prod 148:452–466
182. Alshitawi M, Awbi H, Mahyuddin N (2009) The effect of outdoor conditions and air change rate on particulate matter (PM10) concentration in a classroom equipped with a windcatcher. In: Roomvent
183. Attia S, De Herde AJPOP, Architecture LE (2009) Designing the Malqaf for summer cooling in low-rise housing an experimental study
184. Grosso M, Ahmadi M (2016) Potential cooling energy reduction by a one-channel wind tower: case study modelling in south-Mediterranean climate. Int J Vent 15(3–4):267–287
185. Calautit JK, Hughes BR (2014) Measurement and prediction of the indoor airflow in a room ventilated with a commercial wind tower. Energy Build 84:367–377
186. Calautit JK, O'Connor D, Hughes B (2014) Determining the optimum spacing and arrangement for commercial wind towers for ventilation performance. Build Environ 82:274–287
187. Karakatsanis C, Bahadori MN, Vickery B (1986) Evaluation of pressure coefficients and estimation of air flow rates in buildings employing wind towers. Intersol Eighty Five. Elsevier, pp 769–774
188. Hughes BR, Ghani SAAA (2009) A numerical investigation into the effect of windvent dampers on operating conditions. Build Environ 44(2):237–248
189. Hughes BR, Mak CM (2011) A study of wind and buoyancy driven flows through commercial wind towers. Energy Build 43(7):1784–1791
190. Hughes BR, Ghani SAJIJOSE (2011) A numerical investigation into the feasibility of a passive-assisted natural ventilation stack device 30(4):193–211
191. McCabe C, Roaf SJASR (2013) The wind towers of Bastakiya: assessing the role of the towers in the whole house ventilation system using dynamic thermal modelling 56(2):183–194
192. Ghadiri MH et al (2011) The effect of tower height in square plan wind catcher on its thermal behavior 5(9):381–385
193. Ji Z et al (2012) Performance evaluation and energy saving potential of windcatcher natural ventilation systems in China 1(2):84–95
194. Mohamed M, Nayer A (2019) On wind catcher integration in contemporary buildings in Jeddah. Eqa-Int J Environ Qual 32:1–14
195. Calautit JK, Hughes BR (2016) A passive cooling wind catcher with heat pipe technology: CFD, wind tunnel and field-test analysis. Appl Energy 162:460–471

196. Kolokotroni M, Ayiomamitis A, Ge Y (2002) The suitability of wind driven natural ventilation towers for modern offices in the UK: a case-study. In: World renewable energy congress VII (WREC)
197. Schulze T, Eicker UJE (2013) Controlled natural ventilation for energy efficient buildings. Buildings 56:221–232
198. Chu CR et al (2015) Wind-driven natural ventilation for buildings with two openings on the same external wall. Energy Build 108:365–372
199. Schulze T, Eicker U (2013) Controlled natural ventilation for energy efficient buildings. Energy Build 56:221–232
200. Samuel DGL, Nagendra SMS, Maiya MP (2013) Passive alternatives to mechanical air conditioning of building: a review. Build Environ 66:54–64
201. Ghadikolaei FM, Ossen DR, Mohamed MF (2020) Effects of wing wall at the balcony on the natural ventilation performance in medium-rise residential buildings. J Build Eng 101316
202. Mak CM et al (2007) A numerical simulation of wing walls using computational fluid dynamics. Energy Build 39(9):995–1002
203. Yik FWH, Lun YF (2010) Energy saving by utilizing natural ventilation in public housing in Hong Kong. Indoor Built Environ 19(1):73–87
204. Priyadarsini R, Cheong KW, Wong NH (2004) Enhancement of natural ventilation in high-rise residential buildings using stack system. Energy Build 36(1):61–71
205. Adekoya LJRE (1992) Wind energy end-use: the performance characteristics of a rotating suction cowl 2(4–5):385–389
206. Havens PJEE (2004) A combined fan and wind turbine model for attic turbine ventilators. Healthy Build
207. Lai C-MJE (2003) Experiments on the ventilation efficiency of turbine ventilators used for building and factory ventilation. Buildings 35(9):927–932
208. Rashid D, Ahmed NJWE (2003) Study of aerodynamic forces on a rotating wind driven ventilator 27(1):63–72
209. Lai C-MJE (2006) Prototype development of the rooftop turbine ventilator powered by hybrid wind and photovoltaic energy. Buildings 38(3):174–180
210. Dale JDJAT (1993) Evaluation of the performance of attic turbine ventilators 99(1):14–22
211. Kuo I-S, Lai C-M (2005) Assessment of the potential of roof turbine ventilators for bathroom ventilation. Technology 26(2):173–179
212. Lien J, Ahmed N (2011) Wind driven ventilation for enhanced indoor air quality. In: Chemistry, emission control, radioactive pollution and indoor air quality. IntechOpen
213. Santamouris M, Wouters P (2006) Building ventilation: the state of the art. Routledge
214. Gadi MBJAE (2000) Design and simulation of a new energy conscious system, (ventilation and thermal performance simulation) 65(1–4):355–366
215. Pfeiffer A et al (2008) Modelling of cowl performance in building simulation tools using experimental data and computational fluid dynamics 43(8):1361–1372
216. Geetha N et al (2012) Passive cooling methods for energy efficient buildings with and without thermal energy storage—a review 29(2):913–946
217. Lomas KJ, Cook MJ, Fiala D (2007) Low energy architecture for a severe US climate: design and evaluation of a hybrid ventilation strategy. Energy Build 39(1):32–44
218. Shaviv E, Yezioro A, Capeluto IGJRE (2001) Thermal mass and night ventilation as passive cooling design strategy 24(3–4):445–452
219. Bagiorgas H, Mihalakakou GJRE (2008) Experimental and theoretical investigation of a nocturnal radiator for space cooling 33(6):1220–1227
220. Wang Z, Yi L, Gao FJSE (2009) Night ventilation control strategies in office buildings 83(10):1902–1913
221. Givoni B (1994) Passive low energy cooling of buildings. Wiley
222. Raj VAA, Velraj R (2010) Review on free cooling of buildings using phase change materials. Renew Sustain Energy Rev 14(9):2819–2829
223. Guedes MC (2013) Sustainable Architecture in Africa 421–503

224. Zhang S, Niu JL (2012) Cooling performance of nocturnal radiative cooling combined with microencapsulated phase change material (MPCM) slurry storage. Energy Build 54:122–130
225. Geros V et al (2005) On the cooling potential of night ventilation techniques in the urban environment. Energy Build 37(3):243–257
226. Poirazis H (2004) Double skin façades for office buildings. Report EBD

Chapter 3
Solar Chimney Applications in Buildings

3.1 Introduction to Solar Chimneys

Due to the potential benefits of passive ventilation systems in economic and energy conservation and resistance against noise and carbon dioxide emission [1–3], more research has focused on exploring and improving passive ventilation in the past 20 years. Passive ventilation strategies have been extensively studied over the years. According to local climate conditions and building characteristics, passive ventilation systems show different airflow characteristics and temperature distributions. Simultaneously, some passive ventilation systems also have heat dissipation and heat acquisition functions for space cooling and heating apart from providing adequate ventilation [2, 4–9]. Most modern buildings rely entirely on mechanical ventilation, i.e., active ventilation systems, to satisfy indoor comfort. The majority of the energy supply is used for those active ventilation systems, occupying usable space due to its relatively large volume and structural complexity. Additionally, buildings with mechanical ventilation are often highly airtight to minimize energy consumption and heat loss, resulting in an inadequate fresh air supply [1].

Passive ventilation systems are increasingly being advocated as low-energy alternatives and low-cost solutions for energy conservation buildings. According to the pressure difference sources, typical modes of passive ventilation are referred to as wind-induced, buoyancy-driven, and hybrid ventilation [10, 11]. Corresponding air movement is caused by wind pressure, temperature difference, or both of the above, and humidity difference [12]. It has been found that natural ventilation has the potential to provide adequate capacity for thermal regulation and satisfying indoor air quality in available climatic conditions without reliance on mechanical systems [10, 13, 14].

Passive ventilation systems rely on natural physical mechanisms, which make many uncertainties occur during operation. Wind-induced ventilation systems are solely dependent on prevailing wind speed and incident angle. The stochastics of wind direction and wind intensity bring significant challenges to system performance

L. Shi and H. Zhang, *Solar Chimney Applications in Buildings*, Green Energy and Technology, https://doi.org/10.1007/978-3-031-45218-5_3

evaluation and design [15]. Buoyancy-driven ventilation builds upon the air intensity difference caused by the internal and external temperature difference, ventilating the space even in windless conditions. However, under extremely hot and humid climatic conditions (the temperature difference is insignificant), the system is probably not working properly. Not every passive ventilation system has the dual function of heating and cooling space driven by natural forces. The natural ventilation system can remove the stale warm airflow indoors by accelerating the air movement to provide a space cooling effect. Achieving heating usually requires collecting and storing heat gain and releasing heat when needed to increase the indoor temperature. As the most representative buoyancy ventilation system, the solar chimney has attracted researchers' attention because of its simultaneous ventilation, heating, and cooling functions.

A typical solar chimney is presented in Fig. 3.1. It consists of an absorption wall, a glazing wall, tuyeres/vents, and heat-insulating materials. Airflow is affected by the air density difference between the internal and external environment and the external wind [16–19]. Stale air escapes from the purpose-built openings by the thermo-siphoning effect. The solar chimney components can employ direct or indirect solar energy to drive the airflow in the space. Quesada et al. [20, 21] comprehensively reviewed the research on transparent and translucent solar facades in the past ten years based on theory and experiment and explored its development and applicability. The solar façade absorbs and reflects incident solar radiation and converts direct or indirect solar energy into usable heat. Jiménez-Xamán et al. [22] verified that a rooftop solar chimney applied to a single room for cooling purposes could increase the ventilation rate by 1.16–45.0%. The numerical code was generated to solve the conjugate turbulent heat transfer in a single room equipped with a solar chimney based on the coupling of CFD and global energy balances.

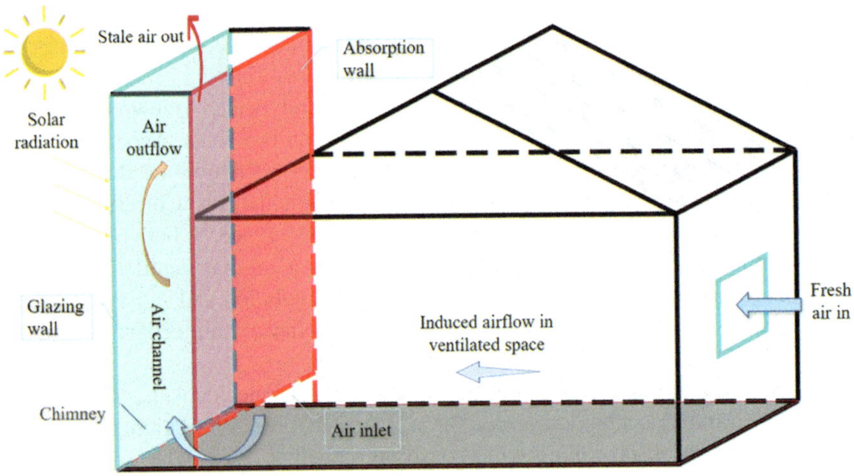

Fig. 3.1 Schematic diagram of a solar chimney

Solar chimneys stand out among many natural ventilation systems not only because of the convenience of their structural features when they are integrated into buildings or in conjunction with other ventilation systems but also because the solar chimney has heating and cooling modes through the cooperation of damping and openings, which makes the structure more sustainable. In order to improve thermal comfort and enhance the applicability of natural ventilation, Monghasemi et al. [4] summarized the existing combined passive ventilation system based on solar chimneys and investigated the thermal regulation of the selected systems and their ability to improve ventilation efficiency.

3.2 Current Research Methods of the Solar Chimney

According to the review of Chen et al. [23, 24], the existing computer-aided prediction models involve numerical models (usually assisted with CFD), analytical models, empirical models, full-scale or small-scale experimental models, multi-zone network models, and zonal models. A considerable part of the research on natural ventilation methods is based on experimental methods coupled with numerical simulations. Because of their complexity and the limitations of modeling methods, most previous studies are without detailed parametric analysis or merely focus on validating analytical models in corresponding experiments. According to the ref. [3], only 31% of the research used analytical methods, 10% of which used pure analytical methods, and the combination of analytical methods and experiments accounted for 13%. The remaining part used analytical methods combined with a numerical value. Based on simplified assumptions, different analytical models are capable of solving a series of energy balance equations under specific scenarios, which in turn would limit the establishment of universal analytical models and solutions.

A prevailing method in passive ventilation uses computational fluid dynamics (CFD) to predict the air distribution and temperature distribution. Using CFD modeling can save time in simulating natural convection and predicting ventilation performance, allowing researchers to quickly develop optimal strategies for improving ventilation performance [25]. CFD can also analyze the multiple flow regimes caused by different driving forces (such as laminar and turbulent) from an intuitive multi-dimensional perspective [26]. With the aids of CFD simulation, Kong et al. [27] determined the optimal inclination angles for the single-chamber roof solar chimney to achieve optimal ventilation performance. Sundar et al. [28] examined eighteen cases under different heat flux intensities and geometrical parameters with the aid of CFD and an experimental method for an inclined solar chimney. Nguyen and Wells [29] used CFD models to predict the ventilation rate and thermal efficiency of wall solar chimneys with four types of adjacent walls, different chimney configurations, and the heat source location. Salari et al. [30] developed a three-dimensional quasi-steady CFD model of a compound solar chimney with the photovoltaic module and phase-change material and verified the three types of combined system performance.

With the aid of CFD-based computational methods, previous studies tried to find the optimal solution to enhance natural ventilation through geometric modification of the chimney configuration and the prediction of the flow pattern in the space connected to the solar chimney. The accuracy of CFD's prediction of the performance of solar chimneys applied to natural ventilation in buildings has been widely recognized. Section 3.3 further discusses the influencing parameters of solar chimney performance in previous studies.

3.3 Influencing Factors of Solar Chimney Performance

Researchers have conducted many parametric studies to find the influencing parameters directly related to the ventilation rate. Shi et al. [31] summarized four groups of influencing factors (a total of thirteen influencing parameters) on solar chimney performance, including the configurations of solar chimney and buildings, installation methods, material properties, and the external environment. Changing the configuration parameters of the system is the most convenient and effective way, which has been reflected in many documents in detail, so this part is not presented in detail here. The influence of external environmental factors is usually related to solar radiation intensity and wind effect. With the continuous emergence of new materials, the properties of different materials also directly affect the performance of the solar chimney. The aggregated influencing factors are listed in Fig. 3.2.

The stack effect intensity is associated with aperture positions and the air density difference between inlet and outlet. Solar radiation intensity as one of the external environmental factors can result in a straight improvement in buoyancy-driven ventilation efficiency without considering the external wind effect. Menchaca-Brandan

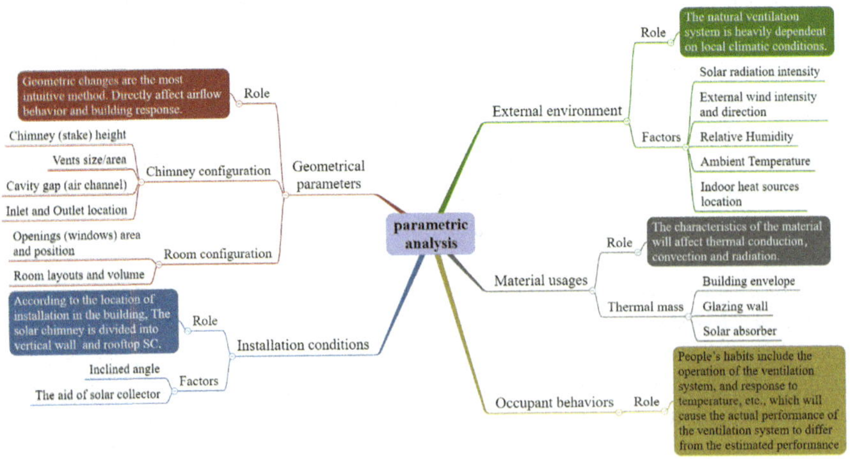

Fig. 3.2 Influencing factors on the performance of the solar chimney

et al. [32] emphasized the importance of considering radiative effects in theoretical models or numerical simulations and experimental setup. The correct establishment of the radiation model will affect the accuracy of the prediction of ventilation efficiency and the accuracy of assessing the thermal comfort of indoor residents. Due to the buoyancy effect brought by the temperature difference, it is known that buoyancy-driven ventilation is preferable in temperate climates [33].

Excepting solar radiation, wind plays an equally critical role in influencing airflow patterns and thermal behavior. Some studies also compared the concurrent influences of solar radiance and external wind on the solar chimney system, and sometimes one driving force could be dominant under a particular condition. Additionally, the effect of wind factors on stack ventilation may be beneficial or unfavorable, depending on the pressure difference caused by the combination of wind pressure and air density difference at the air inlet, whether positive or negative. Shi et al. [34] investigated the interaction between solar chimney performance and external wind numerically and theoretically. When considering the influence of wind, the performance of solar chimneys depends not only on the wind speed but also on the incident angle. Based on the theoretical model developed, the critical wind speed was proposed. Wang et al. [35] verified that the dependence of solar chimney performance and airflow characteristics on solar radiation was reduced when the external wind was taken into account. It was found that keeping the vent size constant under the influence of wind, the optimal design value of the chimney cavity is 0.4–0.5 m, which is different from the previous optimal value of 0.2–0.3 m, which only considers the effect of buoyancy. Even at low wind speeds, solar chimney performance has been enhanced.

The coupling effect of the thermal mass and the buoyancy effect of naturally ventilated buildings has also attracted researchers' attention. Yang and Guo [36] analyzed the nonlinearity of the coupling effect between the thermal mass and the stack effect under an external heat source theoretically. The thermal behavior and ventilation rate fluctuations caused by coupling effects were discussed. Thermal mass-integrated PCM provides an effective method of thermal storage [37]. Through the phase transition, combined with the thermal mass of PCM, naturally ventilated space maintains a relatively uniform temperature. It can significantly reduce the cooling load of the building in mild weather. Especially in the case of night ventilation, the role of PCM becomes more prominent [38, 39]. Vargas-Lopez et al. [40] presented an extensive review of transient mathematical models based on global energy balance models of solar chimneys with/without PCMs. It further developed the mathematical models for a double-channel solar chimney integrated with PCMs.

Figure 3.3 shows the controlling parameters used to evaluate the performance of passive ventilation systems. There are two main parameters used to evaluate and characterize the efficiency of passive ventilation systems from ventilation rate perspectives: air exchange efficiency and ventilation effectiveness [18]. Additionally, thermal efficiency is usually related to indoor temperature regulation and humidity control. For the specific values of each indicator, refer to ref. [41].

Based on the analysis of the influence of control parameters, researchers have carried out extensive research on enhancing stand-alone solar chimney performance. The performance of the solar chimney can be significantly improved by selecting

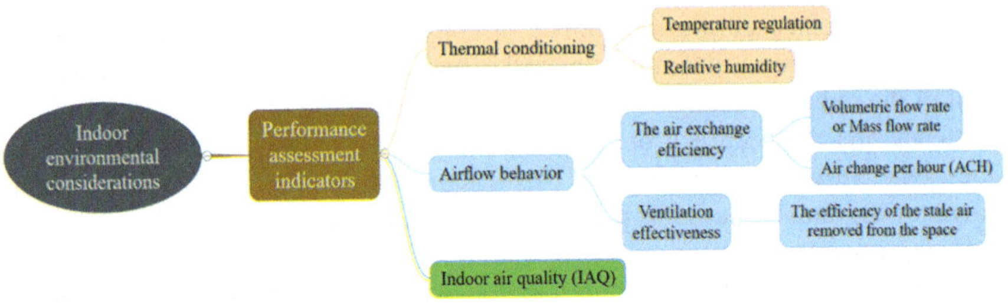

Fig. 3.3 Indicators of performance

the optimal design parameters. Khanal and Lei pointed out [3] that previous studies related to solar chimneys aim to find optimum design solutions for enhancing natural ventilation, considering different design parameters. Most of the research is to optimize the system performance by changing the configuration of the solar chimney components and room openings from the geometric aspect. Enhancing natural ventilation will first consider changing the system's configuration, mainly by changing the geometric parameters to determine the design parameters corresponding to the optimal ventilation efficiency [31]. Secondly, the sensitivity of system performance to external environmental parameters is considered to verify the applicable conditions of the system [17, 31]. Thirdly, based on the analysis of the response, ventilation rate, and thermal behavior of the naturally ventilated building, the desired effect can be achieved by changing the system or building envelope materials, such as solar absorptance of the heat storage materials, heat transmittance, and solar reflectance of glazing cover [17, 42]. Shi et al. [31] comprehensively summarized the correlation between different controlling parameters and solar chimney performance. The diversity of influencing parameters also brings variables to the design of solar chimneys according to local conditions. Zhang and Shi [43] clarified the optimal design value to improve the performance of the solar chimney by reviewing three factors: chimney configurations, installation conditions (chimney installation angle), and material properties.

In addition to geometrically changing the configuration parameters of the solar chimney and the ventilated room, recent studies have shown that some attempts to add components into the chimney cavity can also improve ventilation efficiency. Sheikhnejad et al. [44] employed a passive vortex generator (VG) in the air channel to enhance heat transfer. The effectiveness of this configuration improvement was successfully verified by using simulation technology. Among many studies on improving the heat storage capacity of absorption walls, it is worth mentioning that PCM integration into the solar chimney has been verified to affect ventilation rates and thermal comfort positively. By integrating PCM, a solar chimney can effectively store solar energy during the day and release heat at night, thereby improving the solar chimney performances and making the indoor temperature uniform [45]. Dordelly et al. [46] investigate the influence of integrating a PCM on the performance of two laboratory prototypes of solar chimneys and verified the effectiveness of integrating solar chimney and PCM to improve ventilation efficiency.

Most previous parameterization studies have been based on predicting ventilation rates and building response through geometric modifications. Critical considerations for improving natural ventilation performance include introducing different configurations, such as the inclination position of the rooftop solar chimney, window-to-wall ratio, cavity gaps, stack height, and orifices areas. In addition, there are some studies on the influence of thermal mass and the transmittance and reflectivity of glass materials on ventilation performance. Some parameter analysis also focuses on developing various mathematical models to examine the correlation between airflow or thermal performance and the design variables.

Potential trends and challenges of the solar chimney.

The theoretical studies of solar chimneys in multiple chambers are scarce to support the view that it is feasible to use buoyancy ventilation in buildings with more rooms. Research on the application of independent solar chimneys in a single space has been extensively studied. However, with the increase in volume and structural complexity of modern buildings, the performance of solar chimneys in the multi-chamber is insufficient. The lack of theoretical research has led to the limitations of the solar chimney in the practical application of multi-chamber spaces. The studies on buoyancy-driven ventilation in a multi-chamber building can be divided into two types: the multi-chamber of the single floor and the multi-chamber of the multi-storey building. In recent years, only a few studies have involved the application of solar chimneys in multi-storey buildings. Punyasompun et al. [47] established a three-storey prototype building with a wall-integrated solar chimney and validated the analytical model by experimental data. The optimal structure of the multi-layer solar chimney is verified. That is, only one outlet is set at the top. It was found that the indoor temperature of a multi-storey building using solar chimneys was 4–5 °C lower than that of a building without solar chimneys. By establishing an analytical model and conducting the numerical simulation, Yang and Li [48, 49] compared the ventilation efficiency of a 10-storey building prototype with/without a wall-integrated solar chimney with buoyancy ventilation and mixed ventilation mode and deduced a dimensionless expression and further determined the value of control parameters.

Moreover, the necessary conditions to ensure adequate natural ventilation include natural forces available, building characteristics, exterior envelope air infiltration and airtightness, and occupant behaviors [50–52]. Among them, the envelope's properties with the most direct contact with the outside environment have become the leading research subjects. The coupling effect between the thermal mass of building envelopes and the stack effect has been projected from the improvement methods of the buoyancy-driven ventilation efficiency. However, such coupling effects studies are insufficient for the practice of solar chimneys in previous studies, whether through experimental verification or numerical simulation. Significantly, the practice of solar chimneys in conjunction with different building forms and climatic conditions requires sufficient case studies and theoretical guidance.

Although some studies draw attention to the attempts of combined systems over the years, parameterization research and theoretical analysis are far from enough to meet practical application needs. In a given environment, a coupling of multiple ventilation systems can compensate for each other's performance and increase the service life of a single system. Figure 3.4 shows the independent passive systems combined with solar chimneys based on the selected literature review. Zhang et al. [41] categorized and reviewed the cross-application and combination potentials between independent natural ventilation systems. Compared to the conventional stand-alone passive ventilation system, considerably less research has been conducted on the dynamics and performance of multiple passive-assisted ventilation technologies in sustainable buildings. The study exhaustively recorded considerable attempts in natural ventilation in the past 20 years. Researchers mostly combined solar chimneys with other stand-alone systems and applied them to actual structures in the previous literature.

Based on structural superiority and economic and environmental benefits, the solar chimney has attracted growing interest as an excellent member of passive ventilation systems and advocated for as an auxiliary or alternative system for mechanical ventilation. The current research trend is more focused on combine PV modules with solar chimneys by coordinating components. Sivaram et al. [53] pointed out the commercial applicability of building-integrated passive solar energy technology in which photovoltaic modules and solar still are integrated into a single building configured with a solar chimney. The research of combined systems also shows diversification: the combination of multiple systems to maximize energy saving. Chandavar [54] explored the effectiveness of coupling the solar chimney and PV modules by field tests. As two self-sustained systems, the two components complement each other in their work, thus improving thermal efficiency. With the increasing demand for energy saving, the combination of different ventilation systems is not only limited to double combinations, or triple combinations will get more attempts. Sakhri et al. [55] integrated a windcatcher, solar chimney, and earth-to-air heat exchanger into a test room and conducted a full-scale experiment. Ahmed et al. [13] suggested that a potential solution to combat the heatwaves in warm climates is a combination of a solar chimney, a windcatcher, and evaporative cooling. The related renewable energy systems that could be combined with a solar chimney are listed in Fig. 3.4.

However, the passive ventilation system's performance prediction depends on the local climatic conditions, the experimental settings, and the selection of the prediction method. Therefore, it is challenging and unpredictable to make quantitative comparisons between different passive ventilation combinations. In order to formulate the accessible strategy according to local conditions, not only indispensable theoretical support is needed, but also sufficient case analysis and experimental data are ensured for well-designed combinations, so as to determine the correlation and dependence between different controlling factors and weaken the unexpected correlation of different parameters.

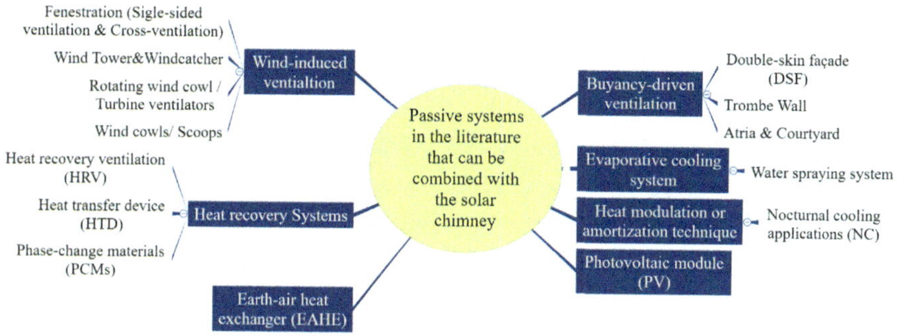

Fig. 3.4 Coupling of passive systems and a solar chimney

3.4 Discussion and Conclusion

It is evident from extensive literature that a solar chimney is an excellent passive ventilation strategy used to enhance natural ventilation and provide thermal comfort. The effectiveness of the solar chimney on the improvement of ventilation rate and thermal regulation has been studied by various researchers numerically and experimentally. Table 3.1 summarizes the ventilation efficiency of independent solar chimneys and passive systems combined with a solar chimney. The indicators for characterizing ventilation efficiency have an extensive range of changes due to different study conditions, such as local climate, experimental settings, prediction tools, and design parameter selection (i.e., geometry modification of solar chimneys and adjacent spaces). It can also be seen from Table 3.1 that the combined system significantly improves temperature regulation and ventilation rate.

An increasing number of studies tend to combine multiple energy-saving systems with SCs with vertical or inclined thermosiphon air channels as an effective passive ventilation enhancement strategy. The strengths of solar chimneys come from providing the desired airflow rate economically, with a simple structure, less space requirement, and ease of integration into existing building facades. Although the research on the application of solar chimneys in single chambers has been going on for a long time, there is no unified theoretical basis and sufficient experimental results for its application in multi-storey buildings. Additionally, the significant differences in the research results in the literature indicate that solar chimneys have not yet been thoroughly studied and developed. More theoretical support and experimental verification are needed to enhance the universality of solar chimneys to be popularized in practical applications. Simultaneously, previous research results have revealed the effectiveness of the multi-purpose application of solar chimneys (i.e., heating and cooling spaces, fire prevention, and smoke control), which also makes the research of solar chimneys very promising.

This article investigates the current development of solar chimneys used in buildings to enhance natural ventilation. The working mechanism and typical structure of

Table 3.1 Performance of a solar chimney and combined systems

Performance	Air temperature adjustment (°C)	Volumetric/mass flow rate (m^3/s, m^3/h, or kg/s)	Air change per hour (ACH)	Energy-saving percentage (%)	Indoor relative humidity (%)
Stand-alone solar chimney	Drop 1.0–5 °C [16, 28, 47, 56–58]; 16.7 °C [59]	50–374 m^3/h [16, 56, 60]; 0.019–0.033 m^3/s [61]; 0.55–44.44 kg/s [22, 28, 62]; 70.6 m^3/h ~1887.6 m^3/h [63]	0.16–15 [22, 56, 57, 60, 64, 65]; 27.11 [59]; 30 [66]	12–50% [63, 67]	
Integrated systems based on solar chimneys	*Drop* 2.0–14 °C with water spaying [57, 68]; 6.7–11.5 °C with evaporative cooling [69–72]; 3.2–9 °C with EAHE [73, 74]; 5.2 °C with windcatcher [75]; 10–13 °C with a windcatcher and EAHE [55]; 8 °C with a Trombe wall and water spraying system [76]; with; *Raises* 14 °C with Trombe wall and PV [77]; 6.4 °C with EAHE [78]	1.4 kg/s with wind tower [79]; 0.0184 m^3/s with EAHE [78]; 0.038–0.144 m^3/s or 130.5 m^3/h or 414 m^3/h with evaporative cooling [69, 71, 72]	35–73 with a wind tower [79]; 12 with PV [80]; 2.42–4.33 with evaporative cooling [70]; 9 with windcatcher [75]	50% with a windcatcher and EAHE [55]; 75–90% with windcatcher [75]	Increases 17% with Trombe wall and water spraying system [76]; increases 28–45% [68]

solar chimneys are briefly presented. The influencing parameters related to ventilation efficiency, temperature distribution, air pattern, and IAQ have been confirmed. This article explores current research gaps of solar chimneys and the various possibilities for enhanced performance. It was observed that a stand-alone solar chimney used in a single room and ideal climatic conditions could effectively provide indoor thermal regulation and enhanced ventilation rate, while a combined system based on solar chimneys can cope with more diverse climatic conditions and provide buildings with functional energy-saving solutions. Previous research has focused on achieving the optimal design strategy based on parameter analysis, including geometrically improving the configuration of the ventilation system and the subsequent response analysis of adjacent spaces. Some of the research hotspots of buoyancy-driven ventilation systems represented by solar chimneys also include the impact of building opening positions on natural ventilation and the coupling effect of the building's thermal mass on buoyancy-driven ventilation. However, since natural ventilation is heavily dependent on local climatic conditions and experimental settings under different working conditions, this brings challenges to quantitative analysis and parallel comparison of system performance. How to improve the stability of the system to cope with different climatic conditions and prolong the durability of the system has become a problem that needs to be solved. In addition, although there have been some attempts at hybrid natural ventilation, the practical application of passive-assisted ventilation systems in a building is limited due to the lack of a large amount of experimental and theoretical support for the coupling system. As the population continues to increase, it is not surprising that multiple rooms have natural ventilation scenarios. At present, only a small amount of research involves the application of solar chimneys in multi-chamber spaces, including single-storey buildings with multi-chambers and multi-storey buildings. However, due to the increase in the amount of spaces, the dimensionality of the variables is undoubtedly increased, thereby increasing the difficulty of theoretical derivation. So far, there is no unified design guide for solar chimneys to enhance natural ventilation in multi-chamber scenarios.

References

1. Geetha N et al (2012) Passive cooling methods for energy efficient buildings with and without thermal energy storage—a review 29(2):913–946
2. Santamouris M, Kolokotsa D (2013) Passive cooling dissipation techniques for buildings and other structures: the state of the art. Energy Build 57:74–94
3. Khanal R, Lei C (2011) Solar chimney—a passive strategy for natural ventilation. Energy Build 43(8):1811–1819
4. Monghasemi N, Vadiee A (2018) A review of solar chimney integrated systems for space heating and cooling application. Renew Sustain Energy Rev 81:2714–2730
5. Zhai XQ, Wang RZ, Dai YJ (2004) Solar chimney combined with underground vent for natural ventilation in energy saving of buildings. In: Proceedings of the 3rd international symposium on heat transfer enhancement and energy conservation, vols 1 and 2, pp 1117–1123

6. Hughes BR, Chaudhry HN, Ghani SA (2011) A review of sustainable cooling technologies in buildings. Renew Sustain Energy Rev 15(6):3112–3120
7. Ascione F (2017) Energy conservation and renewable technologies for buildings to face the impact of the climate change and minimize the use of cooling. Sol Energy 154:34–100
8. Bhamare DK, Rathod MK, Banerjee J (2019) Passive cooling techniques for building and their applicability in different climatic zones—the state of art. Energy Build 198:467–490
9. Sameti M, Kasaeian A (2015) Numerical simulation of combined solar passive heating and radiative cooling for a building. Build Simul 8(3):239–253
10. Chartier Y, Pessoa-Silva C (2009) Natural ventilation for infection control in health-care settings. World Health Organization
11. Awbi H (2010) Basic concepts for natural ventilation of buildings. In: CIBSE BSG seminar: natural and mixed-mode ventilation modelling, University of Reading
12. Etheridge DW, Sandberg M (1996) Building ventilation: theory and measurement, vol 50. Wiley Chichester
13. Ahmed T, Kumar P, Mottet L (2021) Natural ventilation in warm climates: the challenges of thermal comfort, heatwave resilience and indoor air quality. Renew Sustain Energy Rev 138:110669
14. Cheng X et al (2020) Solar chimney in tunnel considering energy-saving and fire safety. Energy 210:118601
15. Horan JM, Finn DPJE (2008) Sensitivity of air change rates in a naturally ventilated atrium space subject to variations in external wind speed and direction. Buildings 40(8):1577–1585
16. Arce J et al (2009) Experimental study for natural ventilation on a solar chimney. Renew Energy 34(12):2928–2934
17. Lee KH, Strand RK (2009) Enhancement of natural ventilation in buildings using a thermal chimney. Energy Build 41(6):615–621
18. Zhai XQ, Song ZP, Wang RZ (2011) A review for the applications of solar chimneys in buildings. Renew Sustain Energy Rev 15(8):3757–3767
19. Shi L et al (2016) Developing an empirical model for roof solar chimney based on experimental data from various test rig. Build Environ 110:115–128
20. Quesada G et al (2012) A comprehensive review of solar facades. Opaque solar facades. Renew Sustain Energy Rev 16(5):2820–2832
21. Quesada G et al (2012) A comprehensive review of solar facades. Transparent and translucent solar facades. Renew Sustain Energy Rev 16(5):2643–2651
22. Jiménez-Xamán C et al (2020) Assessing the thermal performance of a rooftop solar chimney attached to a single room. J Build Eng 31:101380
23. Chen QY (2009) Ventilation performance prediction for buildings: a method overview and recent applications. Build Environ 44(4):848–858
24. Chen Q et al (2010) Ventilation performance prediction for buildings: model assessment. Build Environ 45(2):295–303
25. Jiru TE, Bitsuamlak GT (2010) Application of CFD in modelling wind-induced natural ventilation of buildings—a review. Int J Vent 9(2):131–147
26. Khanal R, Lei CW (2012) Flow reversal effects on buoyancy induced air flow in a solar chimney. Sol Energy 86(9):2783–2794
27. Kong J, Niu J, Lei C (2020) A CFD based approach for determining the optimum inclination angle of a roof-top solar chimney for building ventilation. Sol Energy 198:555–569
28. Sundar S, Prakash D, Surya V (2020) Analysis and optimization of passive wall solar chimney through Taguchi's technique. Appl Solar Energy 56(5):397–403
29. Nguyen Y, Wells J (2020) A numerical study on induced flowrate and thermal efficiency of a solar chimney with horizontal absorber surface for ventilation of buildings. J Build Eng 28:101050
30. Salari A, Ashouri M, Hakkaki-Fard A (2020) On the performance of inclined rooftop solar chimney integrated with photovoltaic module and phase change material: a numerical study. Sol Energy 211:1159–1169

31. Shi L et al (2018) Determining the influencing factors on the performance of solar chimney in buildings. Renew Sustain Energy Rev 88:223–238
32. Menchaca-Brandan MA, Espinosa FAD, Glicksman LR (2017) The influence of radiation heat transfer on the prediction of air flows in rooms under natural ventilation. Energy Build 138:530–538
33. Chan H-Y, Riffat SB, Zhu J (2010) Review of passive solar heating and cooling technologies. Renew Sustain Energy Rev 14(2):781–789
34. Shi L (2019) Impacts of wind on solar chimney performance in a building. Energy 185:55–67
35. Wang Q et al (2020) External wind on the optimum designing parameters of a wall solar chimney in building. Sustain Energy Technol Assess 42:100842
36. Yang D, Guo YH (2016) Fluctuation of natural ventilation induced by nonlinear coupling between buoyancy and thermal mass. Int J Heat Mass Transf 96:218–230
37. Kośny J (2015) PCM-enhanced building components: an application of phase change materials in building envelopes and internal structures. Springer
38. Piselli C et al (2020) Optimal control of natural ventilation as passive cooling strategy for improving the energy performance of building envelope with PCM integration. Renew Energy 162:171–181
39. Prabhakar M et al (2020) Improving the energy efficiency of passive PCM system using controlled natural ventilation. Energy Build 228:110483
40. Vargas-López R et al (2019) Mathematical models of solar chimneys with a phase change material for ventilation of buildings: a review using global energy balance. Energy 170:683–708
41. Zhang H et al (2021) A critical review of combined natural ventilation techniques in sustainable buildings. Renew Sustain Energy Rev 141:110795
42. Barbosa S, Ip K (2014) Perspectives of double skin facades for naturally ventilated buildings: a review. Renew Sustain Energy Rev 40:1019–1029
43. Zhang G, Shi L (2018) Improving the performance of solar chimney by addressing the designing factors. In: IOP conference series: earth and environmental science. IOP Publishing
44. Sheikhnejad Y, Nassab SAG (2021) Enhancement of solar chimney performance by passive vortex generator. Renew Energy 169:437–450
45. Tiji ME et al (2020) A numerical study of a PCM-based passive solar chimney with a finned absorber. J Build Eng 32:101516
46. Dordelly JCF et al (2019) Experimental analysis of a PCM integrated solar chimney under laboratory conditions. Sol Energy 188:1332–1348
47. Punyasompun S et al (2009) Investigation on the application of solar chimney for multi-storey buildings. Renew Energy 34(12):2545–2561
48. Yang D, Li P (2015) Natural ventilation of lower-level floors assisted by the mechanical ventilation of upper-level floors via a stack. Energy Build 92:296–305
49. Yang D, Li P (2015) Dimensionless design approach, applicability and energy performance of stack-based hybrid ventilation for multi-story buildings. Energy 93:128–140
50. Fernandez-Aguera J et al (2019) Thermal comfort and indoor air quality in low-income housing in Spain: the influence of airtightness and occupant behaviour. Energy Build 199:102–114
51. Pomponi F et al (2016) Energy performance of Double-Skin Façades in temperate climates: a systematic review and meta-analysis. Renew Sustain Energy Rev 54:1525–1536
52. Al Assaad D, Ghali K, Ghaddar N (2019) Effect of flow disturbance induced by walking on the performance of personalized ventilation coupled with mixing ventilation. Build Environ 160:106217
53. Sivaram P, Premalatha M, Arunagiri A (2021) Computational studies on the airflow developed by the building-integrated passive solar energy system. J Build Eng 39:102250
54. Chandavar AU (2021) Quantifying the performance advantage of using passive solar air heater with chimney for photovoltaic module cooling. Int J Energy Res 45(2):1576–1586
55. Sakhri N et al (2021) New passive thermal comfort system using three renewable energies: wind catcher, solar chimney and earth to air heat exchanger integrated to real-scale test room in arid region (Experimental study). Int J Energy Res 45(2):2177–2194
56. Bansal N et al (1993) Solar chimney for enhanced stack ventilation 28(3):373–377

57. Chungloo S, Limmeechokchai B (2007) Application of passive cooling systems in the hot and humid climate: the case study of solar chimney and wetted roof in Thailand. Build Environ 42(9):3341–3351
58. Gan GH (2006) Simulation of buoyancy-induced flow in open cavities for natural ventilation. Energy Build 38(5):410–420
59. Villar-Ramos M et al (2020) Parametric analysis of the thermal behavior of a single-channel solar chimney. Sol Energy 209:602–617
60. Khedari J et al (2000) Field measurements of performance of roof solar collector. Energy Build 31(3):171–178
61. Saleem AA et al (2016) Achieving standard natural ventilation rate of dwellings in a hot-arid climate using solar chimney. Energy Build 133:360–370
62. Layeni AT et al (2020) Computational modelling and simulation of the feasibility of a novel dual purpose solar chimney for power generation and passive ventilation in buildings. Sci Afr 8:e00298
63. Zha, X.Y., J. Zhang, and M.H. Qin (2017) Experimental and numerical studies of solar chimney for ventilation in low energy buildings. In: 10th international symposium on heating, ventilation and air conditioning, vol 205. Ishvac2017, pp 1612–1619
64. Mathur J et al (2006) Experimental investigations on solar chimney for room ventilation 80(8):927–935
65. Khedari J et al (2000) Ventilation impact of a solar chimney on indoor temperature fluctuation and air change in a school building 32(1):89–93
66. Imran AA, Jalil JM, Ahmed ST (2015) Induced flow for ventilation and cooling by a solar chimney. Renew Energy 78:236–244
67. Miyazaki T, Akisawa A, Kashiwagi TJRE (2006) The effects of solar chimneys on thermal load mitigation of office buildings under the Japanese climate 31(7):987–1010
68. Rabani R et al (2014) Numerical simulation of an innovated building cooling system with combination of solar chimney and water spraying system. Heat Mass Transf 50(11):1609–1625
69. Swiegers JJ (2015) Inlet and outlet shape design of natural circulation building ventilation systems. Stellenbosch University, Stellenbosch
70. Maerefat M, Haghighi AP (2010) Natural cooling of stand-alone houses using solar chimney and evaporative cooling cavity. Renew Energy 35(9):2040–2052
71. Abdallah ASH et al (2013) Integration of evaporative cooling technique with solar chimney to improve indoor thermal environment in the New Assiut City. Egypt 4(1):45
72. Abdallah ASH et al (2014) Parametric investigation of solar chimney with new cooling tower integrated in a single room for New Assiut city. Egypt Clim 5(2–3):92
73. Ramírez-Dávila L et al (2014) Numerical study of earth-to-air heat exchanger for three different climates 76:238–248
74. Serageldin AA et al (2020) Solar chimney combined with earth to-air heat exchanger for passive cooling of residential buildings in hot areas. Sol Energy 206:145–162
75. Moosavi L et al (2020) New design for solar chimney with integrated windcatcher for space cooling and ventilation. Build Environ 106785
76. Rabani M et al (2015) Empirical investigation of the cooling performance of a new designed Trombe wall in combination with solar chimney and water spraying system. Energy Build 102:45–57
77. Elghamry R, Hassan H (2020) Experimental investigation of building heating and ventilation by using Trombe wall coupled with renewable energy system under semi-arid climate conditions. Sol Energy 201:63–74
78. Elghamry R, Hassan H (2020) Impact a combination of geothermal and solar energy systems on building ventilation, heating and output power: experimental study. Renew Energy 152:1403–1413
79. Bansal NK, Mathur R, Bhandari MS (1994) A study of solar chimney assisted wind tower system for natural ventilation in buildings. Build Environ 29(4):495–500
80. Sivaram P et al (2020) Investigation on a building-integrated passive solar energy technology for air ventilation, clean water and power. Energy Convers Manage 211:112739

Chapter 4
Physical Processes in Natural Ventilation

4.1 Theory of Natural Ventilation

This chapter laid out and amalgamated some essential principles and concepts in detail for completeness. The modeling approach described herein applies the hydrostatic pressure relationship and Bernoulli's theorem along streamlines passing via vents. The modeling approach, including assumptions and approximations, is explained using the development of a mathematical model that depicts the purely natural ventilation of an isolated chamber. Numerous previous researchers have utilized the method of constructing mathematical models to analyze buoyancy-driven ventilation, referring to [1, 2-4].

4.1.1 Conservation Laws

4.1.1.1 Conservation of Mass

The law of conservation of mass tracks the amount of air flowing into and out of a building. Assuming that air is incompressible, the amount of air entering the flow system equals the amount of air leaving.

$$M_{\text{in}} = M_{\text{out}} \tag{4.1}$$

where M stands for the mass dimension per unit time; suppose the rate of change in the air volume due to heating/cooling is minor compared to the overall ventilation volume flow under stable conditions. In that case, the Boussinesq approximation applies, enabling the density of air (ρ) to be viewed as a constant. Consequently, the conservation of mass flux M may be approximated by the conservation of volume flux V, resulting in,

L. Shi and H. Zhang, *Solar Chimney Applications in Buildings*, Green Energy and Technology, https://doi.org/10.1007/978-3-031-45218-5_4

$$V_{\text{in}} = V_{\text{out}};\ M = \rho V \tag{4.2}$$

4.1.1.2 Conservation of Thermal Energy

For a ventilation system where convective heat transfer predominates, conductive and radiative heat transfer elements may be implicitly considered as components of net heat flux from a convective source. Thus, the temperature difference can be determined from the balance between the net heat input (Q_s) from the convective source,

$$Q_s = \rho C_p V \Delta T \tag{4.3}$$

where V is the ventilation volume flux, ΔT denotes the difference in air temperature between the zone's inside and outside, ρ denotes the air density, and C_p stands for the specific heat capacity of the fluid.

4.1.1.3 Conservation of Buoyancy Flux for Buoyancy-Driven Flows

Solar gains and other heat sources generate air density and temperature variations within a building. Its derivation may be explained regarding a Boussinesq flow rising from a heat source $E(W)$. Such a flow is driven by buoyancy flux B $(m^4 \cdot s^{-3})$, which is a function of the reduced gravity g' produced by the source and the associated volume flow rate V $(m \cdot s^{-3})$,

$$B = g'V = \frac{gE}{\rho_0 C_p T_0} \tag{4.4}$$

For a flow in an unstratified environment, there is no removal or addition of heat from or to the buoyant convective parcel as it rises. By conserving thermal energy, E is constant, and the buoyancy flux is conserved concerning height.

If the environment in which the flow develops is stratified, buoyancy flux is conserved only within an environmental layer of the same density. It decreases when the flow ascends into a more buoyant layer because there is less density differential between the flowing fluid and the ambient fluid. By altering the flow's boundary conditions, it is possible to quantify its behavior as it transitions from one layer to another [5].

4.1.2 Auxiliary Equations

4.1.2.1 Variations in Density and Temperature of an Ideal Gas

The ideal gas law (equation of state) establishes the relationship between pressure, density, and temperature. The derived set of formulas requires that the density difference be known. Usually, the known quantities are the net heat input (Q) or the temperatures (T). Also, the relationship between static pressure (p), temperature (T), and density (ρ) connect the heat balance and pressure balance models. Therefore, it is appropriate to set up the relationship between density, temperature difference, and net heat input.

The air can be considered an ideal gas written in the following equation of state:

$$p = \rho R T \tag{4.5a}$$

where R denotes the specific gas constant for air, rewriting Eq. (4.5a) in differential form, Variations in temperature, density, and pressure are interconnected as follows:

$$\frac{dp}{p} = \frac{d\rho}{\rho} + \frac{dT}{T} \tag{4.5b}$$

For natural ventilation flows, the pressure variations are minimal for air to be considered incompressible, so variations in pressure can be neglected. Rewriting Eq. (4.5b),

$$\frac{d\rho}{\rho} = \frac{dT}{T} \tag{4.5c}$$

Integrating Eq. (4.5c), the external ambient air density ρ_0 and temperature T_0 are assumed to be constant and uniform as reference states, resulting in the following density difference:

$$\Delta\rho = \frac{\rho_0 \Delta T}{T_i} = \frac{\rho_i \Delta T}{T_0} \tag{4.5d}$$

where $\Delta\rho = \rho_0 - \rho_i$, $\Delta T = T_i - T_0$; ρ_0 is the ambient air density, ρ_i is the air density inside.

4.1.2.2 Boussinesq Approximation

The conventional Boussinesq approximation is used since it is anticipated that there will not be a significant temperature differential between the solar chimney system's entrance and exit. Typically, the density variations causing stack ventilation are negligible, thus:

$$\frac{\Delta\rho}{\rho_i} \ll 1 \tag{4.6}$$

Under Boussinesq approximation ($\rho_i \approx \rho_0$), rewriting Eq. (4.5b)

$$\frac{\Delta\rho}{\rho_i} = \frac{\Delta T}{T_0} \tag{4.7}$$

4.1.3 Bernoulli's Theorem

Bernoulli's equation restates Newton's Second Law of Motion to a continuously flowing fluid. Based on mechanical energy conservation, a pressure balance along a streamline, Bernoulli's theorem may be used to compute ventilation flow rates. Deriving from the energy conservation principle (appears as a change in kinetic energy), this theorem essentially states that the sum of all forms of mechanical energy along the streamline (a small streamtube with varying cross sections) of a steady flow of an incompressible fluid is constant at all points. For buoyancy-driven flows with uniform density, it can be expressed by,

$$P_0 + \frac{1}{2}\rho u^2 + \Delta\rho g h = \text{constant} \tag{4.8}$$

where ρ denotes the fluid density (kg/m^3), P_0 is an initial pressure (Pa), which is treated as the atmospheric pressure acting on the ambient fluid; u is flow velocity through an orifice (m/s); the second term is fluxes of kinetic pressure, exerted when air through orifices; and the last term is a change in hydrostatic pressure drive due to density deficiency.

The speed u of a flow is related to the volume flow rate V by the general relation,

$$V = C_d A u \tag{4.9}$$

where V represents the volumetric flow rate via the aperture, m^3/s; C_d is the discharge coefficient, and A is the cross-sectional aperture area, m^2.

4.1.4 Flow Through a Vent

Considering airflow through an orifice or duct is involved in ventilating flows analysis. A drop in static pressure across the vent drives the airflow. Figure 4.1 illustrates a single streamline passing through a sharp-edged vent. The airflow converges and accelerates, passing through the plane of vents, and continues to accelerate for a short distance (Z_c) until reaching a maximum uniform velocity at the vena contracta (u_c), where the minimum streamtube cross-sectional area is A_c indicated by horizontal dotted line in Fig. 4.1. Subsequently, the streamlines diverge. The static pressure

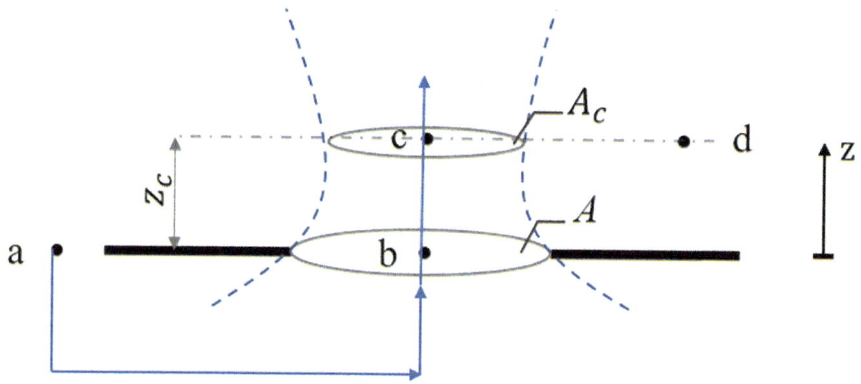

Fig. 4.1. Flow through a horizontal sharp-edged vent with cross-sectional area A. A streamtube denoted by dashed lines confines the flow via the vent

inside the streamtube equals the static pressure outside the streamtube at the vena contracta, i.e., $p_c(t) = p_d(t)$.

Applying Bernoulli's theorem along the streamline across the vent between point a (at $z = 0$) and point d (at $z = z_c$), neglecting the velocity at points a and d away from the vent, i.e., $u_a = u_d = 0$, the fluid velocity at the vena contracta may be linked to the static pressure gradient across the vent.

$$p_a + \rho g z_a + \tfrac{1}{2}\rho u_a^2 = p_c(t) + \rho g z_c + \tfrac{1}{2}\rho u_c^2(t) \quad (4.10)$$

For practical applications, assuming the separation between a vent plane and the vena contracta (z_c) can be neglected, i.e., $z_a \approx z_c$, it is possible to ignore variations in hydrostatic pressure brought on by the vertical height rise between points b and c. As a result, it is possible to quantify the velocity $u_c(t)$ at the vena contracta purely in terms of the disparity between the static pressure across the vent and the air density.

Substituting $p_c = p_d$, and the flow direction is from bottom to top, which assumes that $p_a(t) > p_d(t)$, Eq. (4.10) becomes,

$$\Delta p_{\text{vent}}(t) = p_a(t) - p_d(t) = \tfrac{1}{2}\rho u_c^2(t) \quad (4.11)$$

Assuming a uniform velocity profile across the vent plane and vena contracta, substituting Eq. (4.3–4.9) leads to an expression for the instantaneous volume flux at the vena contracta,

$$V(t) = u_c(t)A_c = C_d A u_c(t) \quad (4.12)$$

By substituting Eqs. (4.12) into (4.11), the volume flux across the vent, *V(t)*, can be in relation to the pressure difference $\Delta p_{\text{vent}}(t)$:

$$\frac{V^2(t)}{2C_d^2A^2} = \frac{|\Delta p_{\text{vent}}(t)|}{\rho} \tag{4.13}$$

This expression can be incorporated into the mathematical model of the pressure drop across horizontal and vertical vents within a ventilated room, which requires the vertical extent of the opening to be much smaller than the floor-to-ceiling height [2, 6].

4.1.5 Hydrostatic Balance

It assumed that fluid velocities are insignificant away from vents, resulting in a hydrostatic pressure distribution such that

$$\frac{dp}{dz} = -\rho g \tag{4.14}$$

where p is the hydrostatic pressure, varying in the z-direction.

Integrating Eq. (4.10),

$$p(z) = -\int_{z1}^{z2} \rho(z) g dz \tag{4.15}$$

The vertical coordinate, z, originates at floor level and is positive upward.

The static pressure in the external environment, room, and chimney cavity, giving,

$$\Delta p = \rho g \Delta h \tag{4.16}$$

where Δp is the pressure difference, Pa; z is Δh is the difference of vertical heights, m.

4.2 Application of the Basic Principles

For example, Fig. 4.2 shows a room ventilated by stack effect only, which will tie together the main points of much of this chapter. The space is solely ventilated through the stack effect. Air travels into and out of the space via flow pathways, including window, floor- and ceiling-level vents. The temperature and density stratification caused by the heat sources within the chimney creates the stack pressure that powers the ventilation flows.

The air inside the storeys gets heat from sun heating. Assuming that the air volume is not confined, raising the air temperature lowers its density; hence, warm air inside a building is less dense than the ambient air outside. This difference in density causes a pressure differential between the air inside and outside the building. In close

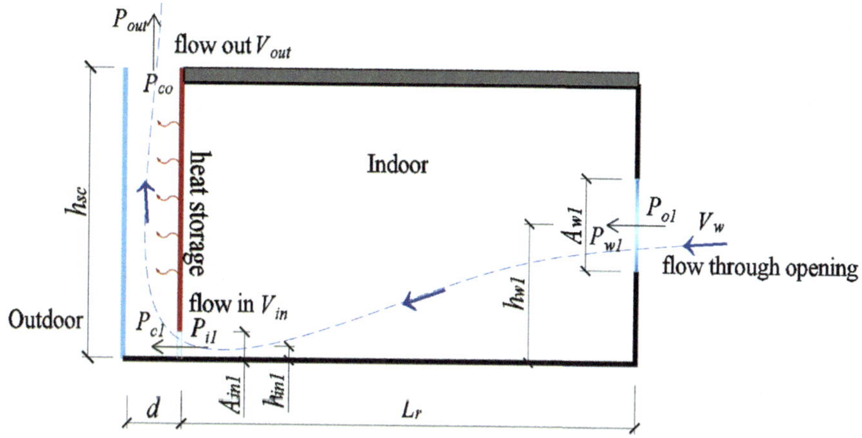

Fig. 4.2. Natural ventilation of side view of a single room through vents by thermal buoyancy. Larger blue arrows indicate the direction of airflow through ventilation openings. Small arrows near the solar chimney absorbing wall represent heat gain from the solar radiation. The sign of the pressure difference is denoted by p, which indicates static pressures. Arrows passing through ventilation vents indicate the direction of the ventilation airflow, which these pressure differences would drive

proximity to the ground, the air pressure is lower than the outside air pressure at the same altitude. This pressure differential causes air to enter the storey through the floor-level vent. The ceiling-level air pressure is greater inside than outside, forcing air to flow through the vent at the ceiling level.

Assume that the room is uniformly heated by a flux source B and that the vents are positioned as shown in Fig. 4.2. At a steady state, the cavity is heated to a temperature T_c while the ambient air temperature is T_0. The room is in hydrostatic balance. Using Bernoulli's theorem combined with the Boussinesq approximation, pressure along the streamline may be traced from the window, the inlet at floor level, to the outlet at ceiling level.

$$\rho_0 u_w^2 + \rho_r u_i^2 + \rho_c u_o^2 = 2g(\rho_0 - \rho_c)(h_{\text{sc}} - h_{\text{inl}}) \tag{4.17}$$

where ρ_0, ρ_r, and ρ_c represent the air density in the ambient, the occupied room, and the chimney cavity; u_w, u_i, and u_o are airflow velocities via the window, air intake, and air exhaust. It effectively states that the total pressure drops across orifices Δp_{vent} (left-hand side) equals the total driving stack pressure Δp_{stack} (right-hand side).

Under steady-state circumstances, achieving mass balance yields the following equation:

$$\rho_0 V_w = \rho_r V_{\text{in}} = \rho_c V_{\text{out}} \tag{4.18}$$

Since the occupied space heat source is not considered, it can be assumed $\rho_r = \rho_0$.

Rearranging the resultant equation Eq. (4.10) with Eq. (4.9), so pressure balance for a single room can be expressed by

$$\rho_c V_{\text{out}}^2 \frac{1}{C_d^2 A_{\text{out}}^2} + \rho_0 V_{\text{in}}^2 \left(\frac{1}{C_d^2 A_{w1}^2} + \frac{1}{C_d^2 A_{\text{in1}}^2} \right) = 2g(\rho_0 - \rho_c)(h_{\text{sc}} - h_{\text{in1}}) \tag{4.19}$$

4.3 Basic Equations of Computational Fluid Dynamics

Computational fluid dynamics (CFD) is a prevalent approach for predicting airflow rate and temperature distribution in passive ventilation. Using CFD-based computational techniques, earlier research attempted to identify the ideal way to improve natural ventilation by modifying the system's geometry and predicting the flow pattern in the region linked to the solar chimney. Using CFD modeling helps expedite the simulation of natural convection and the prediction of ventilation performance, enabling researchers to design practical approaches for enhancing ventilation performance swiftly [7].

CFD relies on the unsteady Navier–Stokes equations (conservation of momentum), continuity equation (conservation of mass), thermal energy equation (conservation of energy), and air equation of state. There are several variants of the equations, depending on the fluid characteristics used. Point-specific instantaneous velocities, pressure, and temperature (or density) may represent the unstable three-dimensional motion of air.

Introduce the fundamental equations of motion, i.e., the Navier–Stokes equations, corresponding to a turbulent flow with non-uniform density (ρ). The viscosity is assumed to be constant, μ_0:

$$\rho \left(\frac{\partial u}{\partial t} + u\frac{\partial u}{\partial x} + v\frac{\partial u}{\partial y} + w\frac{\partial u}{\partial z} \right) = -\frac{\partial P}{\partial x} + \mu_0 \left(\frac{\partial^2 u}{\partial x^2} + \frac{\partial^2 u}{\partial y^2} + \frac{\partial^2 u}{\partial z^2} \right) \tag{4.20}$$

$$\rho \left(\frac{\partial v}{\partial t} + u\frac{\partial v}{\partial x} + v\frac{\partial v}{\partial y} + w\frac{\partial v}{\partial z} \right) = -\frac{\partial P}{\partial y} + \mu_0 \left(\frac{\partial^2 v}{\partial x^2} + \frac{\partial^2 v}{\partial y^2} + \frac{\partial^2 v}{\partial z^2} \right) \tag{4.21}$$

$$\rho \left(\frac{\partial w}{\partial t} + u\frac{\partial w}{\partial x} + v\frac{\partial w}{\partial y} + w\frac{\partial w}{\partial z} \right) = -\rho g - \frac{\partial P}{\partial z} + \mu_0 \left(\frac{\partial^2 w}{\partial x^2} + \frac{\partial^2 w}{\partial y^2} + \frac{\partial^2 w}{\partial z^2} \right) \tag{4.22}$$

where u, v, and w indicate velocity components in the x, y, and z axes, P signifies the pressure at a point. Gravity exerts its influence in the negative z-direction.

Application of the principle of energy conservation to a small control volume yields the following equation for the temperature, T, at a point, representing the equilibrium between convection and conduction of thermal energy:

$$\frac{\partial T}{\partial t} + u\frac{\partial T}{\partial x} + v\frac{\partial T}{\partial y} + w\frac{\partial T}{\partial z} = \frac{k}{\rho C_p} \left(\frac{\partial^2 T}{\partial x^2} + \frac{\partial^2 T}{\partial y^2} + \frac{\partial^2 T}{\partial z^2} \right) \tag{4.23}$$

The mass conservation principle applied to a small element when dealing with the unsteady flow and non-uniform density,

$$\frac{\partial \rho u}{\partial x} + \frac{\partial \rho v}{\partial y} + \frac{\partial \rho w}{\partial z} = 0 \tag{4.24}$$

The Boussinesq approximation, which permits density differences to be overlooked in the N-S equations and affect the acceleration component, may be used to simplify the momentum and continuity equations. Except for the gravitational factor, the density is considered constant ρ_0. Just requiring the inclusion of the buoyancy element. So that Eq. (4.18) can be substituted by,

$$\rho_0\left(\frac{\partial w}{\partial t} + u\frac{\partial w}{\partial x} + v\frac{\partial w}{\partial y} + w\frac{\partial w}{\partial z}\right) = \Delta\rho \cdot g - \frac{\partial p}{\partial y} + \mu_0\left(\frac{\partial^2 w}{\partial x^2} + \frac{\partial^2 w}{\partial y^2} + \frac{\partial^2 w}{\partial z^2}\right) \tag{4.25}$$

where $\Delta\rho$ denotes the density difference $\rho - \rho_0$.

Rewritting Eq. (4.21),

$$\frac{\partial w}{\partial t} + u\frac{\partial w}{\partial x} + v\frac{\partial w}{\partial y} + w\frac{\partial w}{\partial z} = -\frac{\rho}{\rho_0}g - \frac{1}{\rho_0}\frac{\partial P}{\partial z} + \frac{\mu_0}{\rho_0}\left(\frac{\partial^2 w}{\partial x^2} + \frac{\partial^2 w}{\partial y^2} + \frac{\partial^2 w}{\partial z^2}\right) \tag{4.26}$$

* **A note for expert readers**: A large portion of the theory discussed in this chapter is "bookwork," which may be found in many textbooks (such as [8] and [5]) and journal papers (notably [1, 2, 6, 9, 10, 11]).

4.4 Current Work Methods

The majority of the study on natural ventilation systems is based on experimental approaches supplemented with numerical models. Due to the complexity and constraints of modeling approaches, most prior investigations either lacked extensive analytical procedures or only concentrated on verifying analytical models in comparable experiments for simple scenarios involving a single room. A solid validation is required for a complete theoretical model that can be applied in practice. However, in the literature, there is no experiment of steady flows in buildings ventilated by stack effect only for multi-storey wall solar chimneys. Due to the lack of experimental data in a steady state, we can only use relatively complete experimental data of Ref. [12] to verify. Although the model built in this experiment is a wall-mounted solar chimney attached to a single-floor building, the multi-storey building setup in this study uses the same solar chimney form, with windows installed opposite the chimneys on each floor.

Laboratory experiments on large multi-storey solar chimney buildings are rare. The experiments and the simulation approach of the multi-solar chimney that Punyasompun et al. [13] established were non-steady state. Therefore, unsteady experimental data cannot be used to validate this study. We attempted to integrate as much experimental data from the references as feasible for validation purposes. For vertical wall-mounted solar chimneys connected to a multi-storey structure with openings under a steady state, however, relatively little experimental data of steady flows in buildings ventilated by the stack effect are available.

On the other hand, Cheng et al. [14] used the same verification process and comparison method in the research on tunnels with solar chimneys. Based on Ref. [15], they believed that the larger error (i.e., average absolute error of 21.1%) is due to the turbulent airflow inside the room under the air inertia from the window. In the model validation exercise, Cheung and Liu [16] had to build CFD in accordance with the fact that almost all experimental measurements of cross-ventilation in the literature were made on low-rise structures. Therefore, in their research, the experimental results of the previous single-storey building were used to establish a corresponding numerical model for verification, and the reasons for the speed deviation calculated by CFD were discussed. Then the cross-ventilation behavior of high-rise buildings was simulated by CFD. A three-dimensional CFD simulation, which corresponds to the design case of a ten-storey atrium building, is performed to validate the proposed design approach (i.e., a dimensionless theoretical model) in Ref. [17, 18]. Their CFD simulation was carried out by using the software Fluent.

The U.S. National Institute of Standards and Technology created the Fire Dynamics Simulator (FDS), a computational fluid dynamics (CFD) model of fire-driven fluid flow (NIST). A variant of the Navier–Stokes equations suited for low-speed, thermally driven flow is numerically solved by the program, with an emphasis on smoke and heat transfer from fires. Previous research has shown the efficacy and precision of FDS for predicting natural ventilation [19–21]. Those are the reasons why FDS was selected for this study. Fire Dynamics Simulator (FDS) based on large eddy simulation (LES) was adopted to model the prototype structure 1:1. The building opening was set to form the fluid path with natural ventilation. In this study, the energy equations are solved by FDS. Therefore, this work adopts the method to verify the effectiveness of the numerical tools first and then establish the numerical model of the multi-storey building. The research method flowchart is illustrated in Fig. 4.3.

This work determined the validity of the adopted FDS models through comparison with previous experimental data and numerical analysis. The comparisons of the mass flow rate of the numerical simulations and experimental outcomes are detailed in this chapter and elaborated in Ref. [20]. For the 0.1–0.3 m cavity gap scenarios, the comparison error ranges from 0.11 to 13.6 percent, indicating that numerical simulations and experimental results match well. As the cavity gap is expanded, there is a growing discrepancy between the simulation and the experiment (the maximum percentage value of the comparison error is 36.34% at a cavity gap of 1.0 m). The greater the individual variation, the greater the overall variation. The average absolute error is 12.8% for total scenarios. Shi and Zhang [20] noticed that the modeling outcomes deviate slightly from the findings of the experiments. This might result from the testing conditions based on specific test materials and wall thickness with various focuses and limitations.

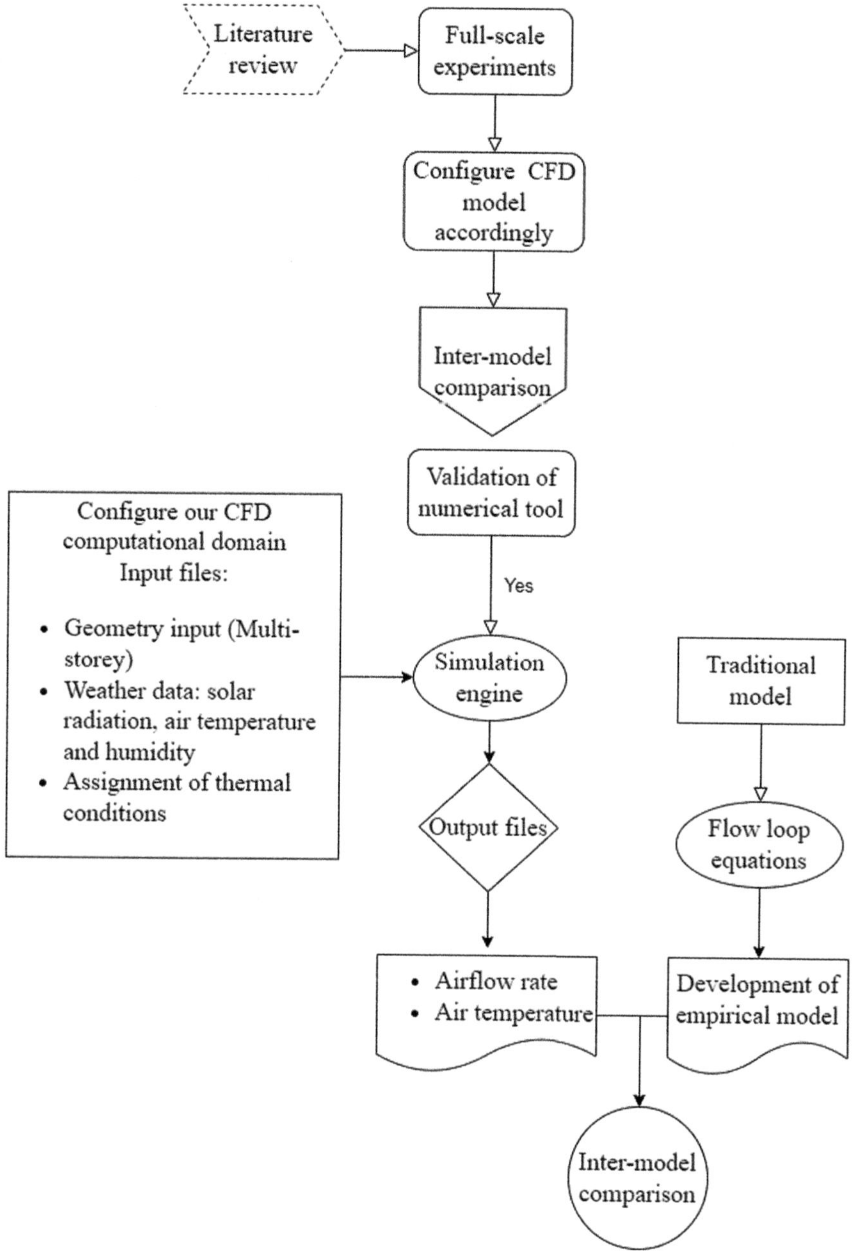

Fig. 4.3. Verification process and comparison method

The data validated above have been included in 'K. McGrattan, S. Hostikka, R. McDermott, J. Floyd, C. Weinschenk, K. Overholt, Fire Dynamics Simulator Technical Reference Guide Volume 3: Validation, National Institute of Standards and Technology, NIST Special Publication 1018–3, 2015' [22]. Besides, Fire Dynamics Simulator (FDS) is a numerical tool developed by the U.S. National Institute of Standards and Technology. FDS model based on LES solves a form of Navier–Stokes equations for thermally driven flow numerically. Numerous validations, calibrations, and investigations on temperature and velocity fields in a fire-induced environment have been performed on the model [23]. Moreover, FDS also has ventilation modules. The mutual verification of ventilation experiments and numerical simulations involving high-rise/tall buildings is relatively common in wind-induced natural ventilation [24-27] utilizing wind tunnel tests. For multi-storey buildings, researchers focus on fire and smoke extraction scenarios for tall buildings more complex than buoyancy ventilation [28]. However, on the basis of a comparison between the numerical prediction with the existing empirical equations and full-scale experimental data, the precision of the numerical scheme of the FDS buoyancy effects was also proved to be credible for buildings and tunnels [19, 29, 30].

4.5 Assumptions and Approximations

For simplicity, a traditional model, Andersen's fully-mixed model [31], has been implemented in the following chapters. We allocate a net heat input to each building zone and adopt a uniform temperature in the traditional model described in our previous study [14]. We tried to analyze the longitudinal linear temperature distribution in the analytical solution. However, due to the complexity of the multi-storey building problem, the mathematical model could never be solved analytically, and it needs to be solved by an iterative algorithm, which makes it impossible to meet the convenience requirements in practical applications. Therefore, we drop the consideration of linear temperature change.

The thoroughly mixed model assumed that the temperature and density of all the air inside the spaces were identically distributed. The air velocities within the building zone are negligible. Any air movement with non-uniform air density is followed by variations in density distribution, leading to a fluctuation in ρ with x and y. There are two density-related simplifications mentioned above; one is the ideal gas law, and the other is the Boussinesq approximation, which is not discussed in depth here.

References

1. Shi L (2018) Theoretical models for wall solar chimney under cooling and heating modes considering room configuration. Energy 165:925–938
2. Connick O (2013) The fluid mechanics of hybrid ventilation. Imperial College London

3. Acred A, Hunt GR (2014) A simplified mathematical approach for modelling stack ventilation in multi-compartment buildings. Build Environ 71:121–130
4. Lin YJP, Huang JJ, Li JS (2013) A study on buoyancy-driven flows in two series-connected chambers. Exp Thermal Fluid Sci 48:37–48
5. Chenvidyakarn T (2013) Buoyancy effects on natural ventilation. Buoyancy Effects Natl Ventilation 1–260
6. Acred A (2014) Natural ventilation in multi-storey buildings: a preliminary design approach. Imperial College London
7. Jiru TE, Bitsuamlak GT (2010) Application of CFD in modelling wind-induced natural ventilation of buildings—a review. Int J Vent 9(2):131–147
8. Etheridge D (2011) Natural ventilation of buildings: theory, measurement and design. Wiley
9. Axley J (1998) Introduction to the design of natural ventilation systems using loop equations. In: Document-air infiltration centre aic proc. Oscar Faber PLC
10. Linden P, Lane-Serff G, Smeed DJJOFM (1990) Emptying filling boxes: the fluid mechanics of natural ventilation. 212:309–335
11. Andersen, K.T., Theoretical considerations on natural ventilation by thermal buoyancy. 1995, American Society of Heating, Refrigerating and Air-Conditioning Engineers ….
12. Bouchair A (1989) Solar induced ventilation in the Algerian and similar climates. University of Leeds
13. Punyasompun S et al (2009) Investigation on the application of solar chimney for multi-storey buildings. Renewable Energy 34(12):2545–2561
14. Cheng X et al (2020) Solar chimney in tunnel considering energy-saving and fire safety. Energy 210:118601
15. Cheng XD et al (2018) Study on optimizing design of solar chimney for natural ventilation and smoke exhaustion. Energy Build 170:145–156
16. Cheung JOP, Liu CH (2011) CFD simulations of natural ventilation behaviour in high-rise buildings in regular and staggered arrangements at various spacings. Energy Build 43(5):1149–1158
17. Yang D, Li P (2015) Natural ventilation of lower-level floors assisted by the mechanical ventilation of upper-level floors via a stack. Energy Build 92:296–305
18. Yang D, Li P (2015) Dimensionless design approach, applicability and energy performance of stack-based hybrid ventilation for multi-story buildings. Energy 93:128–140
19. Jiang Y, Chen QY (2001) Study of natural ventilation in buildings by large eddy simulation. J Wind Eng Ind Aerodyn 89(13):1155–1178
20. Shi L, Zhang G (2016) An empirical model to predict the performance of typical solar chimneys considering both room and cavity configurations. Build Environ 103:250–261
21. Lai CM et al (2013) Determinations of the fire smoke layer height in a naturally ventilated room. Fire Saf J 58:1–14
22. McGrattan K, S.H., McDermott R, Floyd J, Weinschenk C, Overholt K (2020) Fire dynamics simulator technical reference guide volume 3: validation, NIST special publication 1018–3 Sixth Edition. National Institute of Standards and Technology
23. Zhou Y et al (2019) Large Eddy Simulation of effectiveness of solid screen on improving natural ventilation performance in urban tunnels. Tunn Undergr Space Technol 86:174–185
24. Gousseau P, Blocken B, Van Heijst G (2013) Quality assessment of Large-Eddy Simulation of wind flow around a high-rise building: validation and solution verification. Comput Fluids 79:120–133
25. Elshaer A et al (2016) LES evaluation of wind-induced responses for an isolated and a surrounded tall building. Eng Struct 115:179–195
26. Ai ZT, Mak CM (2015) Large-eddy simulation of flow and dispersion around an isolated building: analysis of influencing factors. Comput Fluids 118:89–100
27. Jiang Y, Allocca C, Chen Q (2004) Validation of CFD simulations for natural ventilation. Int J Vent 2(4):359–369
28. Meher S et al (2017) Large Eddy simulation of smoke spread in a multi-storied building. Fluid mechanics and fluid power-contemporary research. Springer, pp 583–593

29. Jiang Y et al (2003) Natural ventilation in buildings: measurement in a wind tunnel and numerical simulation with large-eddy simulation. J Wind Eng Ind Aerodyn 91(3):331–353
30. Durrani F (2013) Using large eddy simulation to model buoyancy-driven natural ventilation. Loughborough University
31. Li YG (2000) Buoyancy-driven natural ventilation in a thermally stratified one-zone building. Build Environ 35(3):207–214

Chapter 5
Theoretical Analysis for a Multi-Zone of the Single Floor

5.1 Introduction

Due to the rapidly rising energy price and operation costs, a growing number of stakeholders in the building sector have shifted their focus on energy supply from conventional to renewable energy resources. Heating, ventilation, and air-conditioning (HVAC) systems are among the biggest energy consumers, which take up 40–60% of the total energy consumption in the building sector [1, 2]. Given the proliferation of studies, a natural ventilation system is usually considered a must-have low-cost solution for sustainable buildings. The natural ventilation performance in energy-saving has been reported in many studies [3, 4, 5].

Natural ventilation systems can provide thermal regulation and improve indoor air quality (IAQ) at a low cost [6, 7]. Compared with traditional mechanical ventilation, those natural ventilation methods have many additional benefits, such as economic ascendancy, encompassing energy consumption mitigation, high durability, and resistance against noise and carbon dioxide emission. It is increasingly advocated as a reliable alternative for existing buildings [8, 9]. One type of natural ventilation strategy makes use of thermal buoyancy, for instance, in solar-driven ventilation systems with open channels. Such systems can be easily integrated into building envelopes [10]. The solar-induced thermal convection and buoyancy may induce airflow circulation due to pressure and temperature gradients throughout the spaces, providing considerable ventilation and thermal comfort to the occupants [11].

The solar chimney has been widely studied and applied in modern buildings as a typical solar-assisted ventilation system. Research aiming to enhance ventilation efficiency and viability of solar chimneys in buildings has been ongoing for decades. Researchers are continually developing mathematical models to address the performance and thermal regulation mechanisms by establishing energy balance equations for various components of natural ventilation systems [12–14]. They have also intensively explored the optimal chimney configuration theoretically and experimentally

L. Shi and H. Zhang, *Solar Chimney Applications in Buildings*, Green Energy and Technology, https://doi.org/10.1007/978-3-031-45218-5_5

[15, 16]. The relationship between air velocity/temperature and the chimney configuration (and the other external variables) is also addressed. The major influencing factors and geometrical parameters have been summarized in the previous study [17].

The studies of solar chimneys mainly involve fluid dynamics and heat transfer in two aspects: the heat transfer process and buoyancy-driven flow mechanics inside the chimney cavity. The other is the response of ventilated spaces, including flow dynamics and thermal behavior [5]. Tominaga and Blocken [18] established five models with different opening locations to study the velocity distribution, the turbulent kinetic energy intensity, and the concentration profiles via wind tunnel experiments. They emphasized the importance of analyzing fluid dispersion and its relationship with the flow pattern determined by a cross-ventilated space's opening position. Besides, as a typical buoyancy-induced ventilation approach, solar chimneys' assessment relies heavily on the natural environment, experimental setup, and prediction methods for parameterization studies. The most prevailing numerical simulation in passive ventilation was carried out with the aid of computational fluid dynamics (CFD) to predict the dynamic characteristics of airflow. CFD can simulate the multiple flow regimes caused by different driving forces from an intuitive multi-dimensional perspective [19]. As one of the CFD models, the effectiveness and universality of the large eddy simulation (LES) model with high temporal and spatial resolution in predicting flow for natural ventilation have been proven [20, 21]. It is found that LES prediction agrees closely with the theoretical estimate [22] and corresponding experimental data in predicting the steady and unsteady airflow and particle trajectories [23-25]. Compared with traditional time-averaged turbulence models, more sophisticated turbulence information can be obtained by solving the time-dependent three-dimensional momentum and continuity of the fluid [26, 27].

The structure of buildings is commonly divided by internal partitions and connected by internal openings. A question then arises about how the interior geometry affects the airstream speed, ventilation efficiency, and flow regime when the building is ventilated by a solar chimney. A practical and essential question is which room/space should be selected to integrate the solar chimney in multi-zone buildings. To date, very few studies have investigated solar chimneys for multi-zone buildings. Existing studies [28–31] have been progressing well in enhancing the ventilation effect of a single space by improving the solar chimney's configuration parameters and thus offering limited information on the inner-flow pattern and airflow distribution in multi-zone spaces ventilated by the solar chimney. Therefore, it is still unclear if solar chimney research findings based on a single-zone system are relevant to multi-zone buildings.

Nevertheless, some existing studies have provided a reasonable basis for solar chimney applications in multi-zone buildings. For example, Lin et al. [14, 32] adopted theoretical and experimental approaches to address the internal geometry's impact on the ventilation flow in two rooms connected via two openings on the partition wall. Punyasompun et al. [16] established two solar chimney types for a three-storey prototype building and validated the model by experimental data. The numerical analysis of the multiple solar chimneys attached to the developed building prototype revealed that the perforated chimney channel is more conducive to ventilation performance. Yang and Li [15, 33] compared three buoyancy-based ventilation schemes via establishing a ten-storey prototype building, including stack-based ventilation with/without an attached chimney and a hybrid ventilation system with the aid of mechanical ventilation. The analytical models validated by numerical simulations were developed, and the corresponding ventilation rates were obtained. The dimensionless expressions of effective vent size were derived based on theoretical analysis. However, the partition (or structure) impacts on the solar chimney performance are still unknown. Compared to conventional single-zone studies, considerably less research has been done on the dynamics of buoyant flows in multiple finite zones considering geometric layouts of the ventilated room.

In this study, the ventilation efficiency and solar chimney's performance attached to a multi-zone building are investigated numerically using an experimentally validated numerical model. The findings of this research provide the basis for practical applications of naturally ventilated buildings. The objectives of this study incorporate the following three key aspects:

- Examine the characteristics (*e.g.*, airflow distribution and ventilation efficiency) of a wall solar chimney attached to a multi-zone building;
- Address the interaction and correlations between the performance (quantified by the airflow rate) and the typical contributing parameters for the wall-integrated solar chimney in the multi-zone building; and
- Propose the optimal design strategies and address the impacts of the structural configurations on the overall ventilation performance of the wall-integrated solar chimney.

5.2 Methodology

5.2.1 Governing Equations

Natural ventilation induced by the buoyancy effect is simulated by the FDS, a numerical tool based on LES developed by the USA National Institute of Standards and Technology. FDS numerically solves the transient Navier–Stokes equations that describe the conservation of mass, momentum, and energy [34]. In the simulation, the geometrical domain is discretized into small control volumes, over which the governing equations are iteratively solved to obtain the flow and thermal characteristics. More details about FDS can be seen in Ref. [35], but the governing equations with the important models are described here for completeness.

The continuity equation is given by,

$$\frac{\partial \rho}{\partial t} + \nabla \cdot \rho \vec{u} = 0 \tag{5.1}$$

The momentum equation is expressed as,

$$\rho\left(\frac{\partial \vec{u}}{\partial t} + (\vec{u} \cdot \nabla)\vec{u}\right) + \nabla p = \rho g + f + \nabla \cdot \vec{\tau} \tag{5.2}$$

The energy equation is written as,

$$\frac{\partial}{\partial t}(\rho h) + \nabla \cdot \rho h \vec{u} = \frac{Dp}{Dt} - \nabla \cdot \boldsymbol{q} + \varepsilon \tag{5.3}$$

where ρ is the density of air, t is the time, $\vec{u}$ is the velocity vector (in x, y, and z directions), p is the pressure, f is the external force vector, $\vec{\tau}$ is the viscous stress tensor, h is the enthalpy, $\nabla \cdot \boldsymbol{q}$ represents the conductive and radiative heat flux, and ε is the dissipation rate at which kinetic energy is transferred to the thermal energy via viscosity of the fluid.

Although the fire dynamics simulator model is commonly used for fire-driven flow, smoke, and fire heat transport, its excellent ability in relation to natural ventilation-related problems has also been demonstrated [36]. FDS can simulate natural ventilation scenarios using computational grids as an efficient flow-solving technique based on LES for steady and unsteady airflows [37], which filters out small eddies while directly solving conservation equations for larger energy-containing eddies [35]. The LES approach accomplishes the scale separation by a low-pass filter of width Δ - the cube root of the cell volume ($\Delta = V_c^{1/3}$, where $V_c = dxdydz$) [38–40], and captures the primary flow features that are larger than the width. For an arbitrary continuous field, Φ, a filtered field is defined as:

$$\Phi(x, y, z, t) = \frac{1}{V_c} \int_{x-\delta x/2}^{x+\delta x/2} \int_{y-\delta y/2}^{y+\delta y/2} \int_{z-\delta z/2}^{z+\delta z/2} \phi(x', y', z', t)\mathrm{d}x'\mathrm{d}y'\mathrm{d}z'. \tag{5.4}$$

This study used Deardorff's model to characterize the net contribution from thermal radiation to the energy equation. The default thermal radiation model that solves the radiation transport equations (RTE) for gray gas [41] is given as follows:

$$s \cdot \nabla I_\lambda(x, s) = \kappa(x, \lambda)[I_b(x) - I(x, s)] \tag{5.5}$$

where $I_\lambda(x, s)$ is the radiation intensity at wavelength λ, $I_b(x)$ is the source term given by the Planck function, s is the unit normal direction vector, and $\kappa(x, \lambda)$ is the absorption coefficient at a point x for the wavelength λ.

5.2.2 Model Description and Boundary Conditions

Figure 5.1 shows that the physical domain considered in this study consists of forced and unforced rooms. The solar chimney shown on the far right consists of an absorption wall and an external glass wall, which form an air/cavity gap. More specifically, there are three connected geometric domains from left to right: unforced room, forced room, and the wall-integrated solar chimney. The forced room is ventilated with the aid of a wall-integrated solar chimney, and the induced airflow ventilates the unforced room. The model also sets the absorption wall's surface to red to distinguish the absorption wall from other walls, as shown in Fig. 7.3a. No external wind is considered in this study.

For the base/reference case (No. 0) of this study, both the forced and unforced rooms have the same dimension of 3.0 m (L_u & L_f) × 3.0 m (W_u & W_f) × 3.0 m (H_R), as shown in Fig. 7.3. The chimney channel has an inlet height $h_{in} = 0.3$ m, a cavity gap $d = 0.3$ m, and a total height of 3.0 m from the floor. The air inlet width is fixed at 3.0 m (equivalent to the reference model's room width). An interconnecting

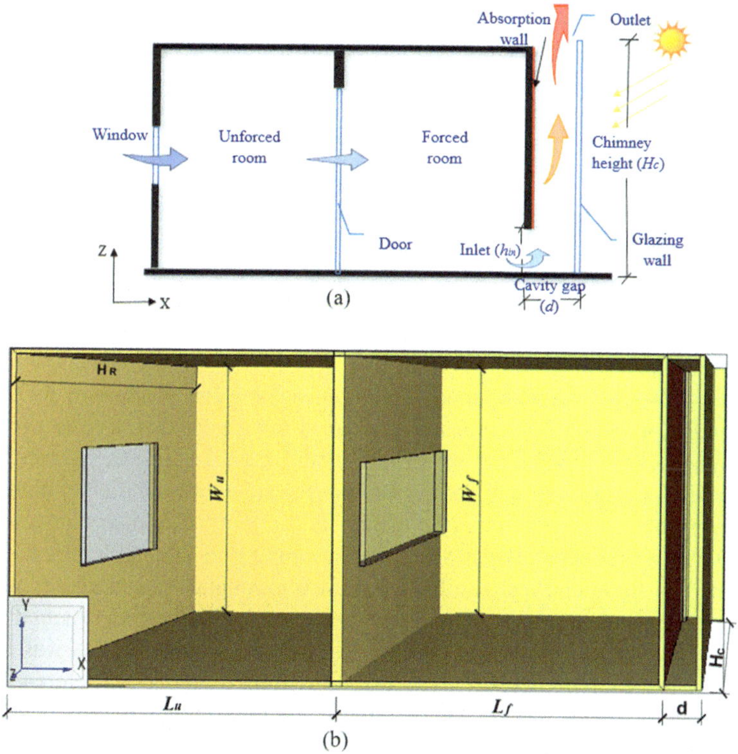

Fig. 5.1 Configuration of the solar chimney and the multi-zone building: **a** an elevation (not to scale); and **b** a top view of a 3D numerical model of the reference case

door with a dimension of 1.0 m (W) × 2.6 m (H) locates at the center of the partition wall between the force and unforced rooms. A window with a dimension of 1.2 m × 1.2 m is located at the center of the unforced room's left wall.

Except for the absorption wall, all the solid surfaces are assumed to be smooth and inert to accelerate the numerical convergency by default. The absorption wall is heated by a constant and uniform heat flux to simulate the solar radiation represented by the red surface attached to the inner wall of the solar chimney. The outside temperature is maintained constant at 20 °C. The emissivity of the glazing wall is set to 0.76 to simulate the transmissivity [42]. The heat transfer coefficient is 15.0 W/(m^2·K).

5.2.3 Validation by Experimental Data

Grid independence of the FDS simulation for a comparable solar chimney configuration has been established in a previous study [39]. Four different grid sizes, namely 0.10 m, 0.08 m, 0.05 m, and 0.04 m, were tested, and grid independence was confirmed with a uniform grid size of 0.05 m. Accordingly, we adopt the same mesh size in the *x*-, *y*-, and *z*-directions in this study, unless specified otherwise. More details about the grid independence test can be found in [39].

Before carrying out the actual numerical calculations, the numerical model based on FDS is validated against the experimental study of Bouchair [43]. The solar chimney and room configurations considered by Bouchair [43] are similar to those shown in Fig. 5.1, but only one room was adopted in the experiments. The absorption and glazing walls were heated by flat heaters, and the temperatures of these walls were adjusted to simulate different solar radiation intensities. Corresponding simulations were carried out under various scenarios, including two air inlet heights (0.1 m and 0.4 m), five cavity gaps (0.1–1.0 m), and four surface temperatures (30, 40, 50, and 60 °C), respectively.

Figure 5.2 shows the comparisons of the mass flow rate of the numerical simulation and experimental outcomes. It is seen in this figure that a good agreement between numerical simulations and experimental outcomes is achieved for the 0.1–0.2 m cavity gap cases (Fig. 5.2a, b). The discrepancy between the simulation and experiment increases with enlarging the cavity gap (with the maximum variations from 39.6% to 57% at cavity gap 1.0 m), as shown in Fig. 5.2e. Although the difference between the numerical results and the experimental data increases with the cavity gap, the numerical results show the same trend with the experimental airflow rates over the surface temperature range. Also, the validation of FDS was effectuated in the previous studies Ref. [42, 44]. According to the above references, it is found that the numerically predicted volume flow rate through the chimney cavity fits reasonably well with experimental data under natural ventilation. The average absolute error is 21.1%. It is known that the consideration of temperature profiles is significantly affecting prediction accuracy. However, there are errors under the extreme cavity gap more significant than 0.5 m and inlet locations, the capability and viability of FDS in predicting buoyancy-driven airflow characteristics, and the solar chimney's overall performance. Hence, FDS is then used to perform the numerical simulations for further numerical scenarios in this study.

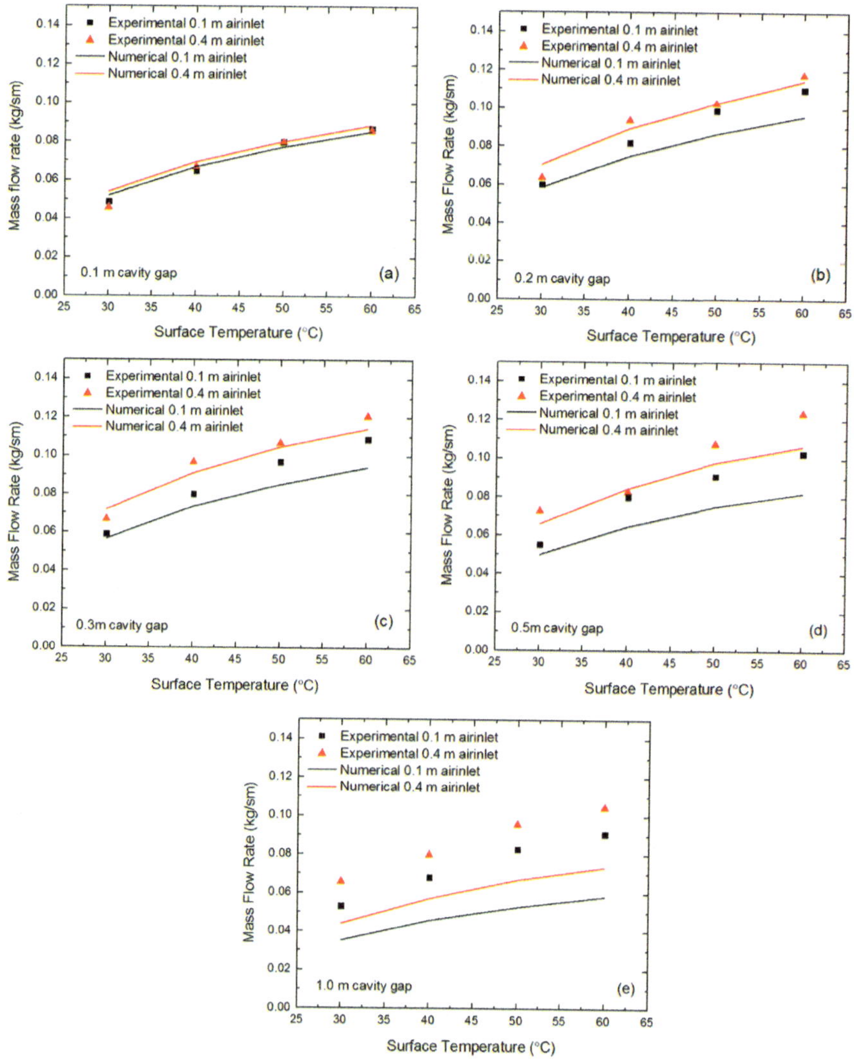

Fig. 5.2 Comparison between numerical and experimental results with different cavity gaps: **a** 0.1 m; **b** 0.2 m; **c** 0.3 m; **d** 0.5 m; and **e** 1.0 m

5.2.4 Simulation Scenarios

Tables 5.1 and 5.2 show the different cases considered in this study to represent specific scenarios. To quantitatively analyze the effect of design parameters, a series of simulations with the means of FDS was carried out systematically. This study quantifies the influence of controlling parameters on ventilation rate, the airflow patterns and velocities inside individual zones, and the air channel associated with buoyancy-driven ventilation. The ventilation rate and airflow patterns are essential indicators of solar chimney performance systematically.

The controlling parameters are primarily obtained from relevant literature and case studies. Four groups of influencing factors on solar chimney performance, including the configurations of solar chimneys and buildings, installation methods, material properties, and the external environment, were summarized in a previous study [17]. According to the objectives of this research and the characteristics of a wall solar chimney, we selected and classified the typical parameters into four groups in this study, including the chimney configuration (e.g., the inlet height and location, chimney dimensions), room configurations (e.g., the internal door position and window area), rooms layouts, and the external environment (e.g., solar radiation).

Table 5.1 A summary of calculated scenarios with various combinations of control parameters

Influencing parameters		No.	Values[a]	Interval	Unit
Base/reference case		0	Forced and unforced room (3 m cube); inlet height: 0.3 m; inlet location[b]: 0 m; cavity gap: 0.3 m; chimney height: 3.0 m; door position[c]: 1.55 m; window area[d]: 1.2 m square; solar intensity: 600 W/m^2	–	–
Chimney	Inlet height (h_{in})	1–4	0.1–0.5	0.1	m
	Inlet location (h_l)	5–8	0–2	0.5	m
	Cavity gap (d)	9–16	0.1, 0.2, 0.3, 0.4, 0.5, 0.8, 1.0, 1.5	–	m
	Height (H_c)	17–20	3.0–5.0	0.5	m
Room	Door position	21–24	0.55–1.55	0.25	m
	Window area (A_{open})	25–28	1 × 1, 1.2 × 1.2, 1.5 × 1.5, 1.8 × 1.8, 2.0 × 2.0	–	m^2
Environment	Solar intensity (q)	29–32	200–1000	200	W/m^2

Note [a]The variations of the different scenarios are based on the reference case. Only one influencing parameter changes each time, while the other parameters are kept the same as those of the reference case; [b]Inlet location represents the vertical distance between the lower edge of the inlet and the floor, [c]Door position refers to the horizontal distance between the centerline of door and wall, where the middle line of the partition is 1.55 m from both sides, and [d]Window is a square of different dimensions located at the center of the wall

Table 5.2 A summary of room layouts with various combinations

No.	Influencing parameters	Unforced room		Forced room		Symbol
		Width[b] (m)	Length[b] (m)	Width[b] (m)	Length[b] (m)	
0	Reference case	3	3	3	3	R0
R1-R5	Unforced room width (W_u)	3–5.5	3	3	3	UW-#[a]
R6-R10	Unforced room length (L_u)	3	3–5.5	3	3	UL-#[a]
R11-R15	Forced room width (W_f)	3	3	3–5.5	3	FW-#[a]
R16-R20	Forced room length (L_f)	3	3	3	3–5.5	FL-#[a]

Note [a] '#' represents the actual value of the parameter under consideration. [b] The room's width and lengths represent the room's dimensions in the Y- and X-directions, as shown in Fig. 7.3b

A total of eight control parameters are considered here. With various planning parameters, 52 numerical scenarios were established by adjusting relevant parameters and comparing results. Twenty-eight scenarios (after deleting the repeated scenarios in Table 7.2) represented the combination of the seven controlling parameters, among which scenario No. 0 is the baseline/reference case.

The impacts of the room configuration on the ventilating airflow were also analyzed by adjusting its side length within a range of 3.0–5.5 m (0.5 m intervals), while the room height in all the cases was kept at 3.0 m. Four influencing parameters were analyzed here, including the widths and lengths of the unforced room (labeled by W_u and L_u) and those of the forced rooms (marked by W_f and L_f). The details of room layouts are listed in Table 5.2. We use 'UW-#' (unforced room width) to denote all cases with different unforced room widths, where '#' represents a value corresponding to the unforced room width. For instance, UW-3 represents the case with an unforced room width of 3.0 m. Similarly, we use 'UL-#' to denote all cases with different unforced room lengths, 'FW-#' for all cases with different forced room widths, and 'FL-#' for all cases with different forced room lengths.

During an FDS calculation, the time step is dynamically adjusted automatically, thereby promoting the solution of the discretized equations to save the computational time cost. The time step size is bound by the convective and diffusive transport speeds within a particular mesh and correspondingly subject to the Courant–Friedrichs–Lewy condition and von Neumann condition. The time steps in each mesh are synchronized during an iteration. The initial time step size is set automatically by dividing the mesh cell size and the characteristic flow velocity. Moreover, the default time step is not allowed to exceed its initial value to avoid numerical instability. The default value of the initial time step size can be specified by $\mathrm{DT} = 5(\delta x \delta y \delta z)^{\frac{1}{3}}/\sqrt{gH}$ s, where δx, δy, δz are the smallest mesh cell dimensions, H is the height of the computational domain, and g represents the acceleration of gravity. Every time step will calculate the number of CFL in each mesh cell, and the time step, δt, is adjusted if the maximum value of the CFL number is out of default values between 0.8 and 1.0. The CFL number is defined by $\mathrm{CFL} = \delta t \cdot \max\left(\frac{|v|}{\delta x}, \frac{|v|}{\delta y}, \frac{|w|}{\delta z}\right) < 1$ where u, v, and w are the velocity components of each mesh cell. Similarly, the von Neumann number is defined by $\mathrm{VN} = 2\max\left(v, D, \frac{k}{\rho C_p}\right)\delta t\left(\frac{1}{\delta x^2} + \frac{1}{\delta y^2} + \frac{1}{\delta z^2}\right) < \frac{1}{2}$ [45].

Furthermore, it is found that the flow time of 150 s is enough to ensure the outputted airflow stabilizes, where the 150 s was then adopted for all the scenarios [42, 46]. The airflow rates through the chimney's air inlet represent the solar chimney's performance for different cases. Thus, the volume flow rates during the last 30 s (between 120 and 150 s of the simulation) were averaged to represent the ventilation capacity and adopted for further analysis in this study.

Besides, the volume flow rate and the indoor flow profiles are examined to evaluate solar chimney performance according to the parameter setting range of this study. Air Change per Hour (ACH) is commonly adopted to indicate air exchange efficiency in practice. It is a measurement of air volume gains or losses per unit volume of space [47], which varies with chimney configurations and solar intensity. Since the ACH heavily depends on the space volume under consideration, the space volume changes from case to case in this study, and the volume flow rate is employed to evaluate the ventilation performance.

5.3 Results and Discussion

5.3.1 *Flow Characteristics*

Solar energy through the glazing wall induces temperature gradient and thermal buoyancy in the vertical chimney channel, thereby driving airflow from the adjacent finite zones to the channel. This section discusses the flow propagation and entrainment of the reference case. Natural convection in the solar chimney cavity heated from the absorption wall promotes buoyant airflow. Individual zones are interconnected by flow paths (i.e., windows, interconnecting doorways, and chimneys) and generate an airflow network. It is evident that a high-velocity region inside the chimney channel and a distinct flow transport induced by buoyancy-driven between the two ventilated spaces connected through a rectangular doorway. A similar flow profile is observed in most scenarios except for some room layouts identified in later sections.

Computational fluid dynamics investigation of air velocity, and buoyancy-driven flows in two series-connected zones equipped with solar chimneys was observed. The induced airflow enters through the window from the unforced room due to the pressure gradient sourced from the buoyancy driving force generated by the solar chimney and escapes from the outlet of a solar chimney attached to the forced room. Figure 5.3 presents the airflow velocity distributions across the mid-planes of the two rooms. The airflow pattern inside the chimney channel is present in all the chimney configurations analyzed in this study. The airstream is accelerated underneath the

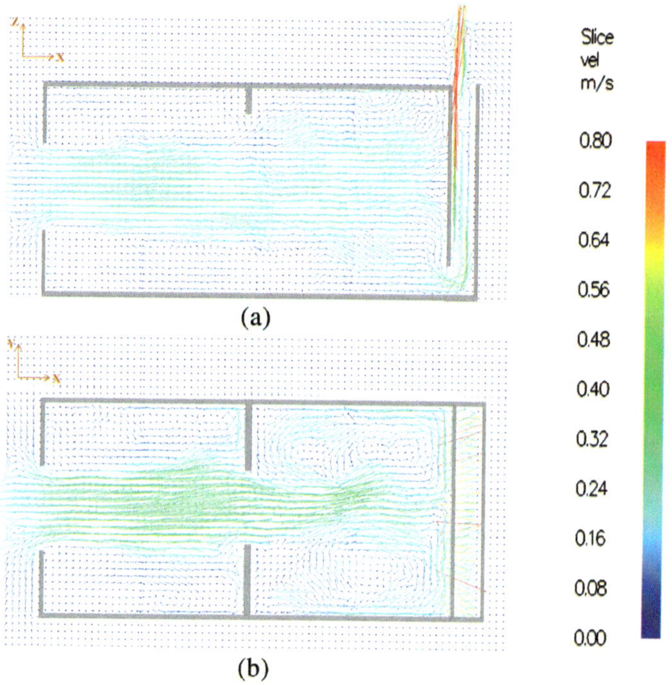

Fig. 5.3 Typical airflow velocity fields on the mid-plane through the two rooms for the reference case: **a** side elevation and **b** top view. The color bar on the right shows a velocity range of 0–0.8 m/s

absorption wall due to thermal buoyancy. Along the absorption wall, the airflow velocities are higher than 0.8 m/s, reaching the maximum at the upper part of the absorption wall. The values near the absorption wall and glazing wall are also higher than in the adjacent two-room since the solar radiation imposed on the absorption wall's surface provides the driving force consistently.

For those models with openings (e.g., window and door) along the center of the space, the related air resistance is relatively more minor and produces fewer vortexes in the rooms. Since the airflow presents a jet through the window, the airflow in the middle can be transferred to the air inlet without obstacles. Under the pressure gradient, a large amount of airflow enters the unforced room through the window. The air forms various zones in the building, such as upward separation, stagnation zone, upward vortex, downward vortex, and flow reversal. It is also can be seen in Fig. 5.3 that the central part of the ventilated zones' velocities is much higher than the regions near the room's sidewalls due to the horizontal motion of the incoming jet.

5.3.2 Chimney Configurations

This section addressed the impacts of the chimney dimensions (i.e., inlet size, cavity gap, chimney height, and inlet location) on solar chimney performance. As shown in Fig. 5.4a, the airflow rates remain approximately constant for the inlet size larger than 0.2 m. The numerical outputs show that the induced airflow volume rate ranges from 0.18 to 0.21 m^3/s. The maximum velocity ranges from 0.75 to 0.85 m/s at the outlet of the chimney. A power regression between volume flow rate and inlet height can be derived, that is, $V = 0.22{h_{in}}^{0.08}$ ($R^2 = 0.5151$), where V is the volume flow rate, and h_{in} represents the inlet height. As the air inlet size increases, the natural flow in the lower part of the room increases, which increases the pressure difference of the chimney effect. However, the growth rate decreases as the air inlet size exceed 0.2 m. Compared with the reference case, the volume flow increased by 1.87% by changing the inlet size to 0.2 m. The optimal air inlet size value is consistent with the literature [48].

Figure 5.4c illustrates the volume flow rate with different cavity gaps. It can be seen that a cavity gap of 0.2 m produces the maximum airflow rate of 0.22 m^3/s among the analyzed cases, which is different from previous studies that showed maximum

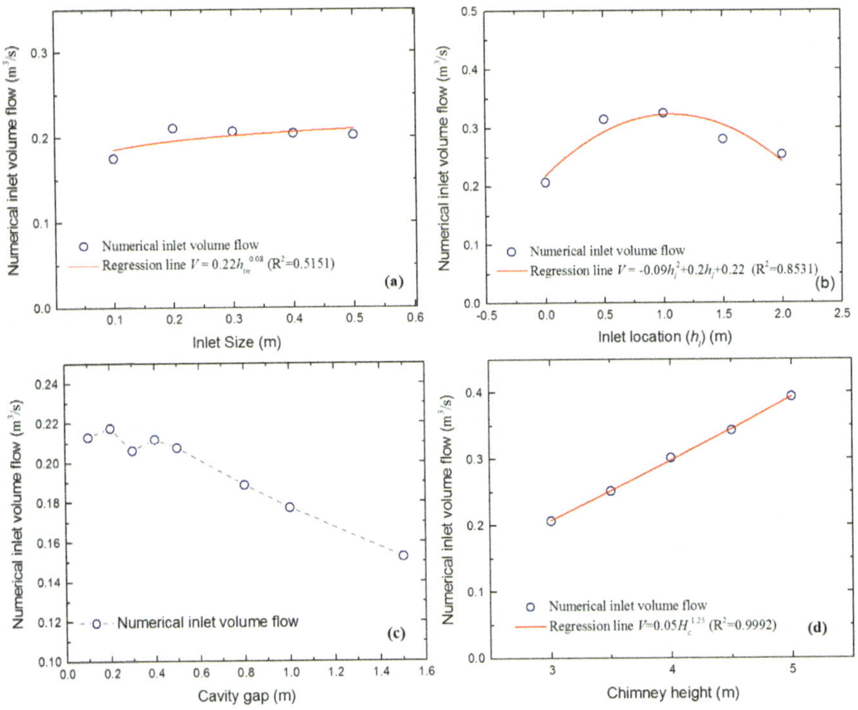

Fig. 5.4 Influences of chimney configuration on the volume flow rate: **a** inlet size; **b** inlet location; **c** cavity gap; and **d** chimney height

performance at the cavity gap of 0.2–0.3 m for the solar chimney connected to a single room [49, 50, 51, 48, 52]. Except for the cavity gap's value that causes the ventilation rate to decrease significantly, the sensitivity of the wall-integrated solar chimney's performance is not significant when the air gap is less than 0.5 m. The volume flow rate remains relatively stable at approximately 0.21 m^3/s with cavity caps ranging from 0.1 m to 0.5 m. As the cavity gap increases beyond 0.5 m, the volume flow rate drops sharply. The relationship between the volume flow rate and cavity gaps (d) expressed by a reported regression in Ref. [17] that $V \propto d^{0.7}$. The results are different within the parameter range of this study. Compared with the benchmark case, the volume flow rate is improved by 5.57% by selecting the optimal cavity gap.

Moreover, the air velocity increases rapidly with the height increase along the absorption wall due to the input heat flux shown in Fig. 5.5. It is observed that velocities near the absorption wall are much higher than those in the central region. Since the temperature rise of the air in the cavity depends on the natural convection process, the air close to the hot wall is first and efficiently heated. Also, the absorbing wall's heating range is limited, and the heating effect becomes weak as the gap increases. Conversely, the velocity gradually decreases as the height increases in the central region and the outer layer. Besides, the reverse flow appears along with the outer layer and the chimney's central area with a larger cavity gap: the larger the cavity gap is, the more pronounced the flow reversal effect will be. The velocity fields reveal that the volume flow rate induced by the heat flux could not continue to grow under a larger cavity gap. The maximum velocity is also affected by the air resistance generated by the reverse flow phenomenon at the air inlet and outlet.

The inlet location, characterized by the distance from the bottom edge of the air inlet to the floor (h_l), was moved upward along the chimney channel's absorption

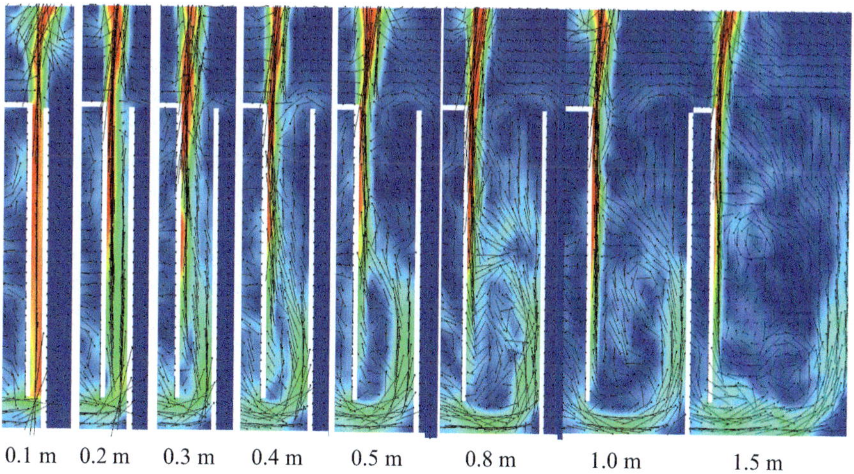

Fig. 5.5 Distribution of the flow transition process in vertical channels with different cavity gaps of 0.1–1.5 m

wall to address its effect on the airflow rate. Figure 5.4b shows that the volume flow initially increases and decreases significantly with the air inlet's continued upward movement, ranging from 0.21 m^3/s to 0.33 m^3/s. The reason for this phenomenon is that the upward movement of the opening leads to the decreased stack effect, which depends mainly on the height difference between the air inlet and the air outlet. As the inlet keeps rising, the effective area with heat inputs is reduced, resulting in a lower airflow rate through the chimney channel. However, the volume flow reaches its peak when the air inlet is at the same level as the window. The incident flow from the window can move to the air inlet with less obstruction and be discharged from the chimney channel, consistent with the previous study [44]. Besides, as the pressure gradient decreases, the maximum velocity of the airstream in the chimney channel decreases from 0.8 m/s to 0.65 m/s. Compared with the reference case, the volume flow can be improved by 57.28% by adjusting the air inlet location.

Similarly, the taller the vertical height is, the more considerable amount of induced air ventilation rate will be. In addition to relatively lower intakes and higher extractors, the stack effect is also favored by increasing chimney heights. As shown in Fig. 5.4d, the volume flow rate rises significantly from 0.21 to 0.39 m^3/s with chimney heights. Comparing to the computational relationship between the volume flow rate and cavity height summarized in Ref. [17]: $V \propto H^{2/3}$, the regression curve shows a similar form of regression: $V = 0.05\, H^{1.25} (R^2 = 0.9992)$. When the chimney height exceeds 4.0 m, which is 1.3 times the floor's height, the maximum flow velocity in the cavity is always maintained at 1.0 m/s. It can be seen that for a multi-chamber building structure, the most effective way to increase the volume flow is to increase the stack height. Compared with the reference case, the volume flow is increased by about 90% by increasing the chimney height to the maximum height selected in this study, namely from 3.0 to 5.0 m.

From the above analysis, the quantitative conclusions of volume flow and chimney configuration can be determined. A relatively optimal cavity gap and inlet height can improve the ventilation rate from the abovementioned discussions. However, when sizes exceed certain values, the improvement becomes minor. If the chimney has a height the same as the building, a preferred design is a solar chimney with a cavity gap of 0.2 m and an inlet height of 0.2 m, with the inlet position located at the floor. The chimney height of 5 m presents the best ventilation efficiency in this study. It was observed from the comparison that the stack height affects the air flow rate most significantly. The location of the air inlet also greatly affects the volume flow rate. Therefore, buildings using the stack effect to enhance the ventilation capacity should adopt tall chimneys and try to make internal/external openings at the same centerline.

5.3.3 *Room Configurations*

With a constant heat flux imposed on the air channel's sidewalls, the door's position does not significantly affect the airflow rate, as illustrated in Fig. 5.6a. As the two connected rooms are purely driven by natural ventilation, the doors are positioned

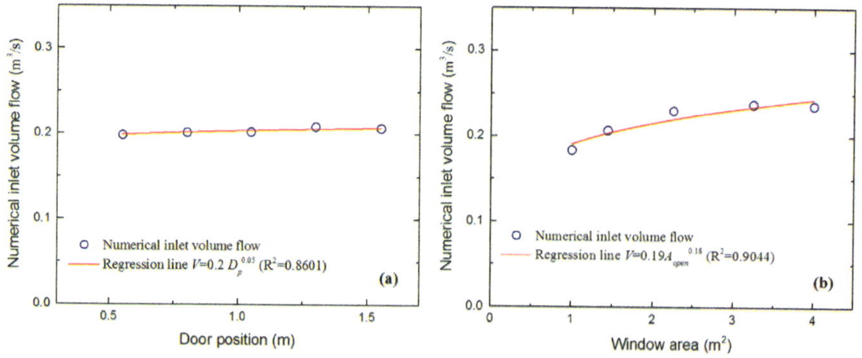

Fig. 5.6 Influence of door position and window area on the inlet volume flow: **a** door position, which refers to the horizontal distance between the center of the door and wall; and **b** window area

opposite the external window to allow the incoming air stream. As the door's position moves from the room's side to the middle, the volume flow rate maintains at about 0.2 m^3/s. Also, as the door position moves from the side to the center of the wall, the volume flow rate increases slightly. Compared with the reference case, the volume flow of the case where the door's position deviates the most from the centerline is reduced by 4.16%.

On the other hand, the volume flow rate increases significantly with the window area, as seen in Fig. 5.6b. The regression curve shows a fit power growth trend $V = 0.19A_{\text{open}}^{0.18}$ ($R^2 = 0.9044$). A reported regression in Ref. [17] shows $V = A_{\text{open}}/(0.65 + 5.2A_{\text{open}})$, where A_{open} is the area of the window connected to the outside environment. Compared with the empirical equation in the literature, the performance of the multi-zone solar chimney is more sensitive to the size of the opening connecting the building to the outside. Larger openings undoubtedly increase the air supply, which in turn affects the volume flow at the air inlet of a solar chimney. In this group, the volume flow rate ranges from 0.18 to 0.24 m^3/s as the window area increases from 1 m^2 to 4 m^2. The maximum velocity in the cavity remains 0.8 m/s. Compared with the reference case, with the increase of the window area (within the value range of this study), the volume flow rate increases by about 14.98%.

Despite the minimal effect on the flow rate, the door position significantly affects the velocity distribution and flow patterns. Figure 5.7 shows the velocity contours and the formation of vortices under various door positions. Door positions that deviate from the airflow path generate more resistance in the central part and harm velocity distribution. The boundary layer over the internal partition wall occupies a large proportion of the cross section. Hence, most of the flow is influenced by the boundary obstruction, which produces a significant portion of flow swirls and eddies from its regular path. Such an uneven airflow distribution is likely to cause local discomfort. Conversely, when the window is directly opposite the door, a relatively more minor flow is restricted due to the boundary layer, which results in a somewhat higher air velocity.

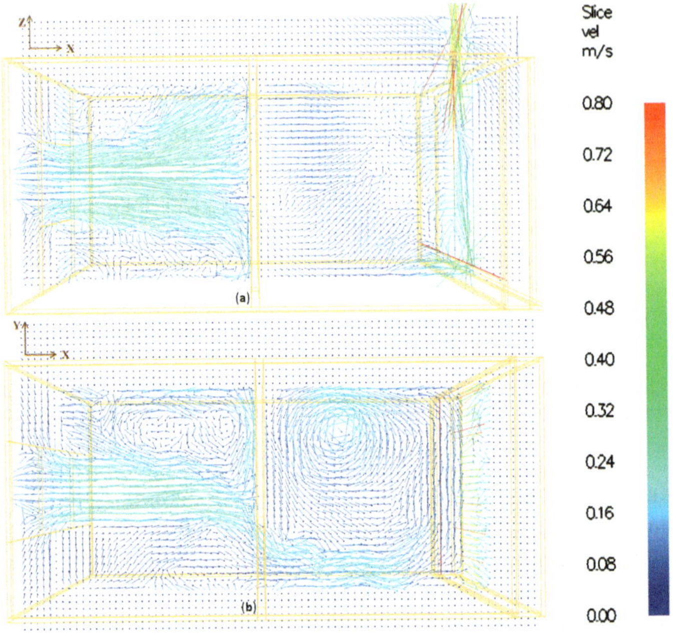

Fig. 5.7 Velocity fields of ventilated zones with the door position of 0.55 m at 140 s; **a** velocity vector slice of elevation; **b** velocity vector slice of top plan view; the color bar on the right shows a velocity range of 0–0.4 m/s

The numerical study on the volume flow rate in two series-connected rooms for different aspect ratio are studied through over 20 cases. The airflow patterns of buoyant flows in multiple chambers/zones with varying ratios of aspect were investigated. Figure 5.8 shows the changes in volume flow rates. An insignificant difference in the airflow rate was noticed as the room size increased accordingly. The volume flow rates in different room layouts fluctuate slightly in all room cases, ranging from 0.2 m^3/s (scenario R9) to 0.21 m^3/s (scenario R11). The differences in volume flow in each group (six test scenarios) are barely noticeable. Comparing the reference case, it can be seen from the four room configurations that the volume flow variation of each group does not exceed 5% by changing room dimensions: the volume flow rate of the unforced room width group (R1-R5) increases by 2.09%, and the mean value is 0.21 m^3/s; the unforced room length group (R6-R10) increases 2.68%, and the mean value is 0.208 m^3/s; the volume flow rate of the forced room width group (R11-R15) increases 2.73%, and the mean value is 0.209 m^3/s; the volume flow rate of the forced room length group (R16-R20) increases 2.71%, and the mean value is 0.21 m^3/s. The difference between the maximum and minimum volume flow rates is 4.09% within all the room configurations. Thus, changing the layout of those rooms to improve volume flow is not as effective as the other factors, such as the chimney configuration and window area.

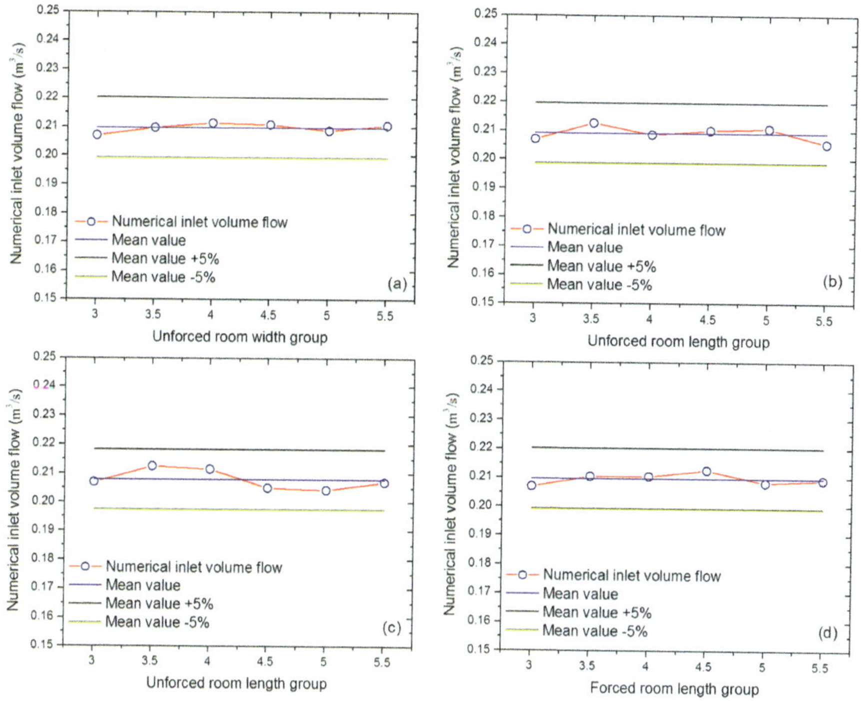

Fig. 5.8 Influence of room layout on the volume flow rate

Despite the minor improvement in ventilation rates, different layouts significantly impact the indoor velocity, turbulent fluctuations, and flow patterns, such as the recirculation area, rotation features, and vortex center. The effect of room layout on ventilation performance is analyzed by airflow pattern and velocity fields at two slices positions (Planes Y and Z), taking cases of (R4, (3, 3) (L) × (5, 3) (W)) and (R14, (3, 3) (L) × (3, 5) (W)) as an example. The velocity contours and vectors, along with the horizontal plane Z and vertical plane Y for both cases at 140 s, are presented in Fig. 5.9. The diagrams show how air movement changes according to different room combinations. The indoor air velocity in the central region of the room is much more significant than the average velocity along the walls. The total volume of the rooms remains unchanged, and the openings of the rooms remain constant, while the connections of different rooms make different flow patterns and velocity distributions in space after experiencing contraction and expansion through the spaces.

In the case of R14 (see Fig. 5.9c, d), when the unforced room is relatively more extensive, the airflow produces more separation and vortex in the large room. The fresh air supply has a limited penetration depth into the large room and generates a uniform mixing through the room due to the boundary layer's effect and the moving air. Most airflows are restricted, resulting in low air velocity. The flow velocity of most areas in the unforced room is below 0.16 m/s, while the airflow along the center

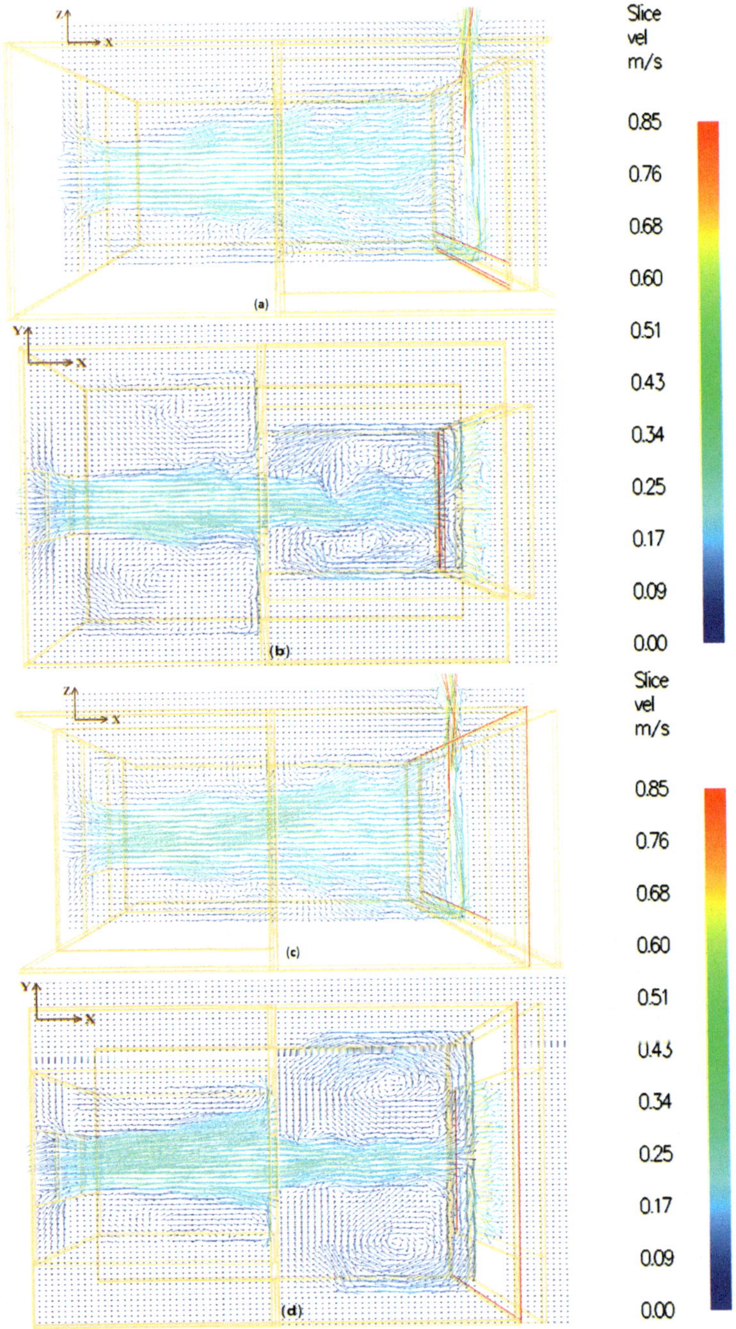

Fig. 5.9 Velocity field in unforced/forced room width change group: **a** vertical velocity field of case R4; **b** plan velocity field of case R4; **c** vertical velocity field of case R14; **d** plan velocity field of case R14. The color bars on the right show a velocity range of 0–0.85 m/s

of the room from the window is maintained above 0.3 m/s. In contrast, case R4 (see Fig. 5.9a, b) presents different velocity distributions, mainly at the central location of the jet, and two separate air flows on either side of the jet in the small room. Moreover, when space suddenly shrinks, relatively faster airflow is observed in the small room, while the air in most parts of the large room has smaller and weaker swirls, except for the central area of the room. Airflow forms a larger stagnant area in the unforced room.

It can be seen from Fig. 5.10 that rooms with great width differences give worse ventilation performance compared to cases of changing the magnitude in the length direction due to deep airflow penetration and unobstructed airstream. Two-room cases are chosen as examples for illustration. Figure 5.10 compares the velocity contours of two representative cases (R9, (5,3) (L) × (3,3) (W)) and (R19, (3,5) (L) × (3,3) (W)) at two planes (Planes Y and Z) at 140 s. For the R19, it can be seen that when the forced room is set as a large room, the airflow produces more separation and a fast vortex in both chambers. Most of the air flows faster due to the jet passing directly toward the solar chimney. For the R9, it is shown that limited flow penetration into the forced room generates an uneven mixing through the room. In the cases of R9 (seen Fig. 5.10a, b), air movement is noticeable in the forced room. The airflow through the unforced chamber encounters the space suddenly shrinks, more fast airflow is observed in the small room. However, it can be observed that there are stagnant areas in the upper and lower parts of the unforced room.

It is worth noting that although R9 and R19 are long and narrow structures, it can be seen from the color bar that different combinations could affect the airflow speed within the rooms and chimney cavity. When the side length of the forced room is shorter than that of the unforced room, it will slightly drop the flow velocity in the rooms and the chimney cavity. For case R9, the unforced room's side length is longer than that of the forced room; that is, the large room is directly connected to the outside world. It can be seen from Fig. 5.9a, b that the indoor retained air increases, and the flow velocity is slightly less than in the former case. This is due to the increased frictional resistance due to long-distance flow and convection when the air flows through the unforced room.

By comparing the velocity fields of the ventilated internal area with Figs. 5.9 and 5.10, the more considerable the width difference between the two rooms, the more sizeable the stagnant area is. Although a shorter path is more conducive to generating high-speed airflow due to the loss during air transportation, too wide a room structure makes the flow rate not evenly distributed. It could also restrict the airflow due to the obstruction of the wall.

5.3.4 External Environment

Solar-based natural ventilation system highly relies on the intensity of solar radiation. In the benchmark case, solar heat flux intensity varies from 200 to 1000 W/m^2, and accordingly, the volume flow rate changes from 0.351 to 0.592 m^3/s at ambient

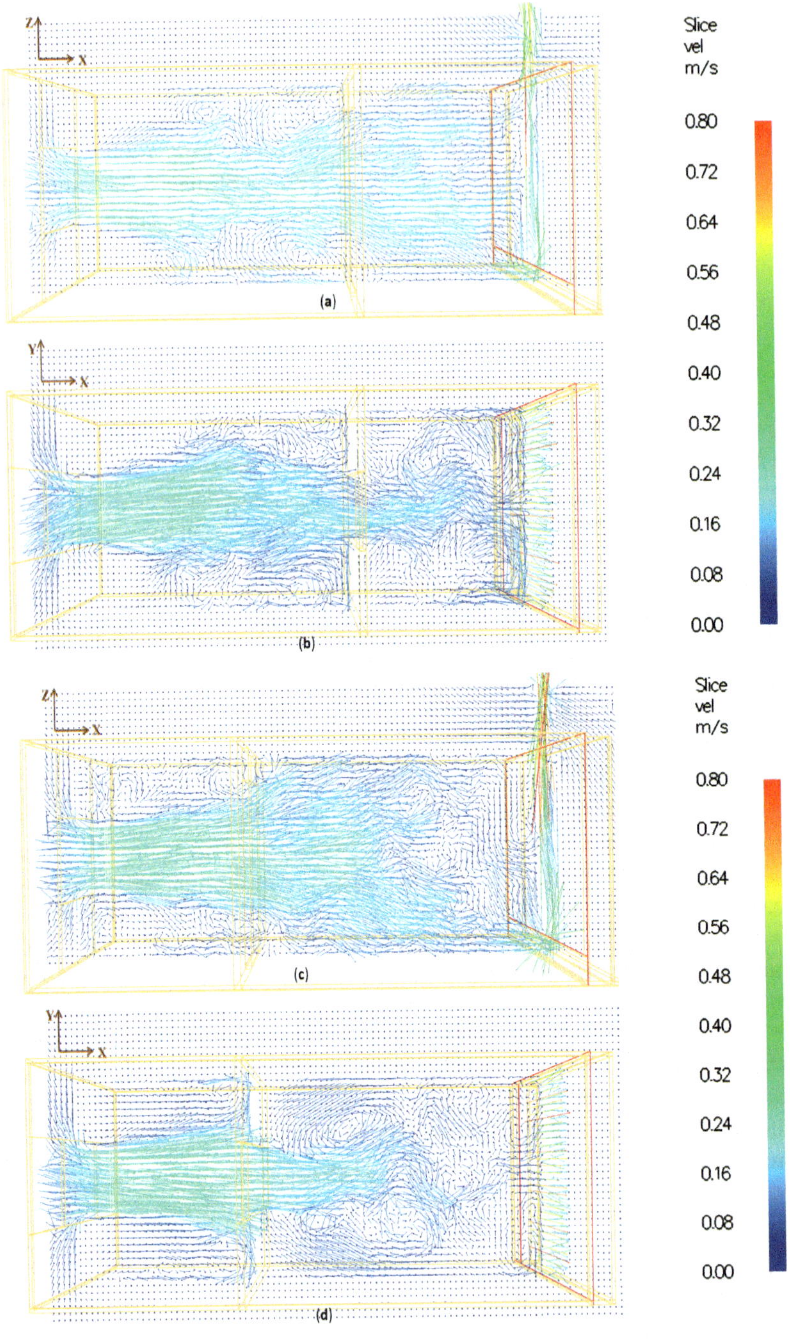

Fig. 5.10 Velocity field in unforced/forced room length change group: **a** vertical velocity field of case R9; **b** plan velocity field of case R9; **c** vertical velocity field of case R19; **d** plan velocity field of case R19. The color bars on the right show a velocity range of 0–0.8 m/s

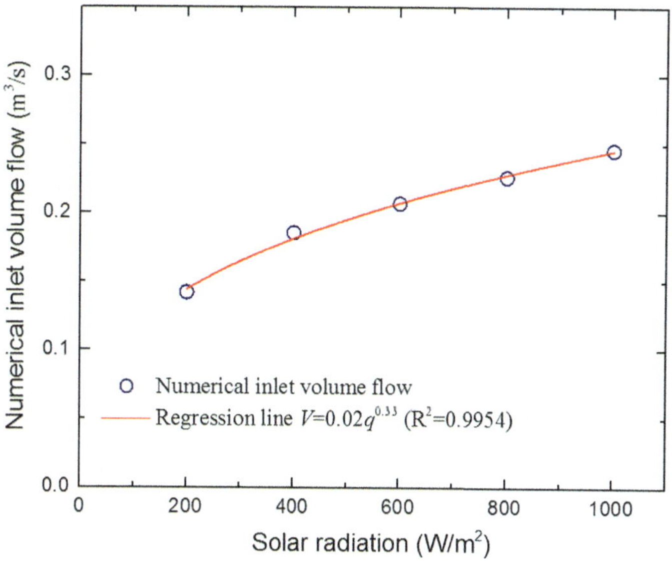

Fig. 5.11 Influence of heat input intensity on air inlet flow

temperatures (20 °C), as shown in Fig. 5.11. A regression expression $V = 0.02q^{0.33}$ ($R^2 = 0.9954$) describes the relationship between volume flow rate and solar radiation intensity. Comparing to the similar computational relationship summarized in Ref. [17]: $V \propto q^{1/2}$ and $V \propto Q^{0.572}$, where q is the heat input intensity, W/m^2; Q is the heat input, W, there is a slight reduction in power compared with single-room cases. The corresponding value increases almost linearly.

Similarly, the maximum velocity in the cavity increases from 0.55 to 0.95 m/s. The driving force of the solar-assisted ventilation system is to use solar radiation to raise the temperature in the cavity to form a phenomenon of buoyancy as the air velocity is increased under relatively higher thermal buoyancy. Therefore, the higher the solar radiation is, the more significant the temperature difference will be—high-intensity solar radiation results in better ventilation performance. Compared with the reference case, the volume flow increases by 18.9% as the solar radiation increases from 600 to 1000 w/m^2.

5.4 Recommendations and Implementations

A series of scenarios were investigated numerically with various control parameters for two connected spaces ventilated by a solar chimney under various chimney configurations and room layouts. This study extends the numerical model of a single-zone building in the infinite homogeneous environment presented in previous studies to the cases of multiply connected spaces of a single floor. The mathematical model will be utilized to predict the behavior of the airflow in buildings through purpose-built openings with different influential variables. The potential correlation between geometrical parameters and ventilation efficiency can be addressed for the two series-connected rooms in this study, which were compared and validated with existing experimental data in Ref. [17, 53, 54-56]. Like those single-zone applications, the main influential factors of solar chimney performance under multi-zone building applications still include chimney configuration, opening size, and solar radiation. The configuration of the two rooms (i.e., width and length) and the position of the middle door show limited impacts on the airflow rate. However, they significantly affect the indoor airflow regime and velocity distributions, which should be considered in the related thermal comfort studies.

Table 5.3 shows the research outcomes of this study, partially agreeing with the results developed in previous studies with single-zone applications. It can be seen that the empirical correlations between the cavity gap decreased for the two-room with a single solar chimney. In contrast, the impact of chimney height on ventilation performance was consistent with that of a single chamber. At the same time, the chimney height and inlet location more directly affect the volume flow rate. The influence of solar radiation intensity is similar to the case of single-zone applications. Figure 5.12 reflects the percentage improvement of each parameter to the ventilation rate according to the results mentioned above to present the controlling parameters that impacted the ventilation rate intuitively.

The optimal design parameters can be found by analyzing the volume flow between the control group of chimney configurations. Based on the analyzed scenarios, the optimal solar chimney has an inlet size of 0.2 m and a cavity gap of 0.2 m for a series-connected space for all the scenarios (Table 7.2 and Table 5.2). Improvement seems insignificant beyond the proposed values (i.e., inlet size and

Table 5.3 A comparison of empirical models in this study and literature

Factors	Multi-zone building in this study	Single-zone building in the literature
Inlet size	$V \propto h_{\text{in}}^{0.08}$	$V =$ $0.16 - 0.009A_{\text{in}}/A_{\text{out}} - 0.025A_{\text{out}}/A_{\text{in}}$ [39]
Inlet location	$V = 0.22 + 0.2h_l - 0.09{h_l}^2$	–
Cavity gap	–	$V \propto d^x$, $x \in [0.652, 0.756]$ [53–55]
Stack height	$V \propto H^{1.25}$	$V \propto H^x$, $x \in [0.539, 0.667]$ [54-56]
Solar radiation	$V \propto q^{0.33}$	$V \propto q^x$, $x \in [0.452, 0.500]$ 54-56
Door position	$V \propto D_p^{0.05}$	–
Window area	$V \propto A_{\text{open}}^{0.18}$	$V = A_{\text{open}}/\left(0.65 + 5.2A_{\text{open}}\right)$ [17]

Note V is the volumetric flow rate; H_l is the inlet location, the vertical distance between the inlet center and the floor; H is the height of the chimney cavity; q is the solar radiation; D_p is the door position that, is the horizontal distance between the centerline of door and wall, where the middle line of the partition is 1.55 m from both sides; A_{open} is the area of window openings; A_{in} is the inlet area, and A_{out} is the outlet area

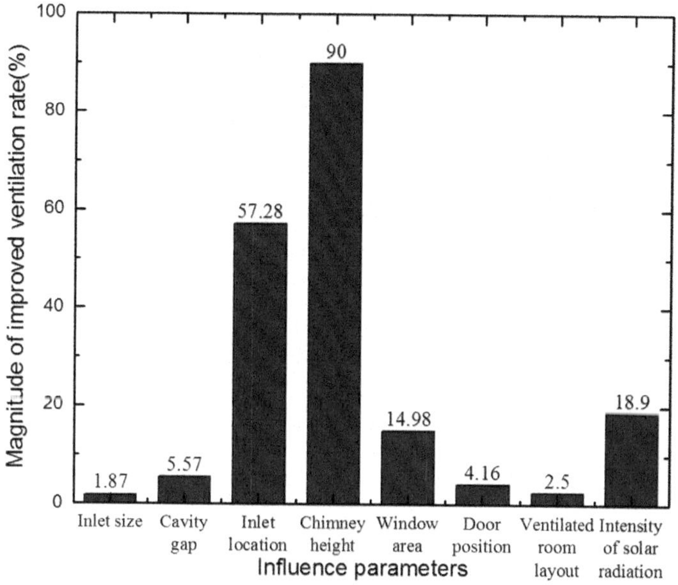

Fig. 5.12 Sensitivity of volumetric flow to each influencing parameter

cavity gap). Besides, better ventilation performance can be achieved when the air inlet is in the middle of the room.

Moreover, for solar-assisted natural ventilation in multi-zone buildings, increasing the flow rate is necessary to enhance the solar chimney performance to drive the airflow in a larger space without increasing the number of solar chimneys. Two prerequisites are generally needed to improve stack ventilation, i.e., relatively higher extractors and sufficient heat sources (internal or external heat sources). The intensity of the stack effect is associated with the height difference between air intake and export and the density difference of the internal and external air. Therefore, increasing the chimney height to the maximum is preferred when the construction conditions and standards allow. Furthermore, as increasing intensive solar radiation, buoyancy provides adequate driving forces to extract the buildings' stagnant air, causing the airflow through the chimney channel to accelerate and improve the volume flow rate profoundly.

According to the parameter range calculated in this study, the main parameter of the room layout that affects the ventilation rate is the window opening area. After weighing the airflow distribution and volume flow rate, the interior door position should be centered on the same axis as the window to improve solar chimney performance. As the openings' size increases, more air is drawn into space, which significantly increases the volume flow rate. However, the volume flow rate generated by the solar chimney in different room layouts is maintained at about 0.21 m^3/s. It can be seen from the velocity field distribution that the long and narrow combination is better than a wide room shape. The more extensive boundary layer interacts with the airflow, and a large part of the airflow is obstructed, and backflow occurs. After weighing the airflow distribution and volume flow rate, the interior door position should be centered on the same axis as the window.

Furthermore, naturally ventilated buildings require to be relatively narrow to allow the airflow to penetrate deeply into a building. The long and narrow room shape is more conducive to ventilation by comparing the volume flow rate and velocity profiles. By weighing airstream speed against the flow patterns, it presents a sound ventilation effect when the side length of the forced room (i.e., the room directly connected to the solar chimney) is more extended than that of the unforced room.

5.5 Conclusions

This study numerically investigated the solar chimney installed for a series-connected space and addressed the influences of room layout and spatial structure on natural ventilation performance. Focusing on connected rooms, which is less discussed in previous solar chimney studies, comprehensive influencing factors were investigated, and critical contribution parameters in a two-room scenario were identified. The concluded empirical models for the two-room structures will improve solar chimney designs in versatile building structures.

The numerical prediction of flow patterns and volume flow rate within the building showed that the chimney configuration and solar radiation intensity could significantly improve the ventilation rate than the room configuration (i.e., width and lengths). However, the room configuration significantly affects the velocity field and the airflow pattern related to the thermal comfort level, which should be considered in the related thermal comfort studies.

Under analyzed conditions, an increase in stack height and window area can result in a straight improvement in the ventilation rate. There are optimal design parameters in the cavity gap and inlet height when the chimney height is constant. The improvement is significant before 0.2 m, but if it exceeds the critical value, the influence on the flow rate is insignificant, and the growth rate gradually decreases. Furthermore, the building layouts and geometries show limited impact on the ventilation performance, while the flow pattern and velocity distribution are sensitive to the building layouts and change significantly.

References

1. Pérez-Lombard L, Ortiz J, Pout C (2008) A review on buildings energy consumption information. Energy and Buildings 40(3):394–398
2. Orme MJE, Buildings (2001) Estimates of the energy impact of ventilation and associated financial expenditures 33(3):199–205
3. Geetha N et al (2012) Passive cooling methods for energy efficient buildings with and without thermal energy storage–a review 29(2):913–946

4. Santamouris M, Kolokotsa D (2013) Passive cooling dissipation techniques for buildings and other structures: the state of the art. Energy Build 57:74–94
5. Khanal R, Lei C (2011) Solar chimney—a passive strategy for natural ventilation. Energy Build 43(8):1811–1819
6. Cuce E et al (2019) Sustainable ventilation strategies in buildings: CFD research. Sustain Energy Technol Assess 36:100540
7. She F et al (2019) Sustainable energy saving alternatives in small buildings. Sustain Energy Technol Assess 32:92–99
8. Shi L et al (2018) Determining the influencing factors on the performance of solar chimney in buildings. Renew Sustain Energy Rev 88:223–238
9. Zhai XQ, Song ZP, Wang RZ (2011) A review for the applications of solar chimneys in buildings. Renew Sustain Energy Rev 15(8):3757–3767
10. Monghasemi N, Vadiee A (2018) A review of solar chimney integrated systems for space heating and cooling application. Renew Sustain Energy Rev 81:2714–2730
11. Teshnehdel S et al (2020) Data-driven and numerical approaches to predict thermal comfort in traditional courtyards. Sustainable Energy Technol Assess 37:100569
12. Vargas-López R et al (2019) Mathematical models of solar chimneys with a phase change material for ventilation of buildings: A review using global energy balance. Energy 170:683–708
13. Shi L (2018) Theoretical models for wall solar chimney under cooling and heating modes considering room configuration. Energy 165:925–938
14. Lin YJP, Linden PF (2002) Buoyancy-driven ventilation between two chambers. J Fluid Mech 463:293–312
15. Yang D, Li P (2015) Natural ventilation of lower-level floors assisted by the mechanical ventilation of upper-level floors via a stack. Energy and Buildings 92:296–305
16. Punyasompun S et al (2009) Investigation on the application of solar chimney for multi-storey buildings. Renewable Energy 34(12):2545–2561
17. Shi L et al (2018) Determining the influencing factors on the performance of solar chimney in buildings. Renew Sustain Energy Rev 88:223-238
18. Tominaga Y, Blocken B (2016) Wind tunnel analysis of flow and dispersion in cross-ventilated isolated buildings: Impact of opening positions. J Wind Eng Ind Aerodyn 155:74–88
19. Khanal R, Lei CW (2012) Flow reversal effects on buoyancy induced air flow in a solar chimney. Sol Energy 86(9):2783–2794
20. Zhang X, Weerasuriya AU, Tse KT (2020) CFD simulation of natural ventilation of a generic building in various incident wind directions: Comparison of turbulence modelling, evaluation methods, and ventilation mechanisms. Energy and Buildings 229:110516
21. Durrani F, Using large eddy simulation to model buoyancy-driven natural ventilation. Loughborough University
22. Abdalla IE et al (2007) Large-eddy simulation of buoyancy-driven natural ventilation in an enclosure with a point heat source. Int J Comput Fluid Dyn 21(5–6):231–245
23. Beghein C, Jiang Y, Chen QY (2005) Using large eddy simulation to study particle motions in a room. Indoor Air 15(4):281–290
24. Jiang Y, Su M, Chen Q (2003) Using large eddy simulation to study airflows in and around buildings. ASHRAE Trans 109(2):517–526
25. Durrani F, Cook MJ, McGuirk JJ (2015) Evaluation of LES and RANS CFD modelling of multiple steady states in natural ventilation. Build Environ 92:167–181
26. Davidson L, Nielsen PV (1996) Large eddy simulations of the flow in a three-dimensional ventilated room
27. Yang T et al (2006) A comparison of CFD and full-scale measurements for analysis of natural ventilation. Int J Vent 4(4):337–348
28. Wang QY et al (2020) External wind on the optimum designing parameters of a wall solar chimney in building. Sustain Energy Technol Assess 42:1–10
29. Al-Kayiem HH, Sreejaya KV, Chikere AO (2018) Experimental and numerical analysis of the influence of inlet configuration on the performance of a roof top solar chimney. Energy Build 159:89–98

30. Rattanongphisat W, Imkong P, Khunkong S (2017) An experimental investigation on the square steel solar chimney for building ventilation application. Energy Procedia 138:1165–1170
31. Gorjian S et al (2021) Recent advances in net-zero energy greenhouses and adapted thermal energy storage systems. Sustain Energy Technol Assess 43:100940
32. Pourghorban A, Asoodeh H (2022) The impacts of advanced glazing units on annual performance of the Trombe wall systems in cold climates. Sustain Energy Technol Assess 51:101983
33. Yang D, Li P (2015) Dimensionless design approach, applicability and energy performance of stack-based hybrid ventilation for multi-story buildings. Energy 93:128–140
34. Li Y, Heiselberg PJIJOV (2003) Analysis methods for natural and hybrid ventilation-a critical literature review and recent developments 1(4):3–20
35. McGrattan K et al (2013) Fire dynamics simulator technical reference guide volume 1: mathematical model 1018(1):175
36. Jiang Y, Chen QY (2001) Study of natural ventilation in buildings by large eddy simulation. J Wind Eng Ind Aerodyn 89(13):1155–1178
37. Jiang Y, Allocca C, Chen Q (2004) Validation of CFD simulations for natural ventilation. Int J Vent 2(4):359–369
38. Zhai ZJ et al (2007) Evaluation of various turbulence models in predicting airflow and turbulence in enclosed environments by CFD: part 1—summary of prevalent turbulence models 13(6):853–870
39. Shi L, Zhang GM (2016) An empirical model to predict the performance of typical solar chimneys considering both room and cavity configurations. Build Environ 103:250–261
40. Adamski MJB, environment (2007) Optimization of the form of a building on an oval base 42(4):1632–1643
41. Kämpf JH, Robinson DJE, Buildings (2010) Optimisation of building form for solar energy utilisation using constrained evolutionary algorithms 42(6):807–814
42. Shi L (2019) Impacts of wind on solar chimney performance in a building. Energy 185:55–67
43. Bouchair A (1989) Solar induced ventilation in the Algerian and similar climates. University of Leeds
44. Shi L et al (2019) Interaction effect of room opening and air inlet on solar chimney performance. Appl Therm Eng 159:113877–113888
45. McGrattan KB et al (2000) Fire dynamics simulator—technical reference guide. National Institute of Standards and Technology, Building and Fire Research
46. Shi L, Zhang G (2016) An empirical model to predict the performance of typical solar chimneys considering both room and cavity configurations. Build Environ 103:250–261
47. Ashrae A (2016) Ventilation and acceptable indoor air quality in low-rise residential buildings
48. Bassiouny R, Koura NSA (2008) An analytical and numerical study of solar chimney use for room natural ventilation. Energy Build 40(5):865–873
49. Arce J et al (2009) Experimental study for natural ventilation on a solar chimney. Renew Energy 34(12):2928–2934
50. Miyazaki T, Akisawa A, Kashiwagi TJRE (2006) The effects of solar chimneys on thermal load mitigation of office buildings under the Japanese climate 31(7):987–1010
51. Park D, Battaglia F (2015) Application of a wall-solar chimney for passive ventilation of dwellings. J Solar Energy Eng-Trans ASME 137(6)
52. Gan GH (2010) Simulation of buoyancy-driven natural ventilation of buildings-Impact of computational domain. Energy Build 42(8):1290–1300
53. Burek SAM, Habeb A (2007) Air flow and thermal efficiency characteristics in solar chimneys and Trombe Walls. Energy Build 39(2):128–135
54. Ryan D, Burek SAM (2010) Experimental study of the influence of collector height on the steady state performance of a passive solar air heater. Sol Energy 84(9):1676–1684
55. Ryan D, Burek S, Baker P (2005) Experimental investigation into the buoyancy driven convection in passive solar heating facades. In: The 2nd Scottish conference for postgraduate researchers of the built and natural environment. Glasgow Caledonian University, UK

56. Shi L et al (2016) Developing an empirical model for roof solar chimney based on experimental data from various test rigs. Build Environ 110:115–128

Chapter 6
A Mathematical Framework of Modeling SC-Assisted Stack Ventilation

6.1 Introduction

Overusing traditional energy resources has led to many practical problems, such as energy crisis and environmental pollution. The building sector accounts for roughly 40% of total energy usage for space cooling and heating [1]. Renewable energy resources have been popularly adopted in those applications of energy-efficient buildings to mitigate the energy burden. The use of active/passive solar-based technologies for heating/cooling, thermal storage, and power generation purposes attracts the attention of many researchers [2–4]. The efficiency of such systems depends on the geometric parameters of solar-based systems [5, 6] and the improvement of solar collector thermal performance through the use of novel materials, such as filled graphene nanoplatelets nanofluid [7–9] and phase-change materials [10–14], to maximize solar radiation collection and conversion into usable heat energy and transfer to a working fluid [15, 16]. Significantly, curtailed ventilation consumption may help reduce peak electricity consumption and gas emissions in various residential and commercial applications [17]. They apply natural physical processes to promote air movement and achieve a satisfactory indoor environment. Passive ventilation designs with air layers could achieve multiple functions following the airflow pattern, including natural ventilation, space heating/cooling, and preheating air [18].

Solar chimneys (SCs) are attractive in energy-saving buildings as a reliable passive ventilation system to amplify natural ventilation. Buoyancy-driven natural ventilation ventilates the building even without wind, based on the difference in air density due to internal and external temperature differences. The induced airflow is accelerated within an openable air channel and the adjacent occupant spaces using solar-induced thermal convection, conduction, and radiation to generate the stack effect. Maghrabie et al. [19] comprehensively reviewed SC-based strategies for building heating and cooling and combined systems based on SCs to enhance system performance in unavailable climates to achieve low-energy buildings ultimately. The previous research attempted to determine the geometric parameters of

L. Shi and H. Zhang, *Solar Chimney Applications in Buildings*, Green Energy and Technology, https://doi.org/10.1007/978-3-031-45218-5_6

SCs to maximize the ventilation rate in the chimney cavity or naturally ventilated buildings [20]. The most prevalent parameterization studies related to the solar-induced fluid movement were conducted using computational fluid dynamics (CFD) to model the various flow regimes and thermal behaviors and solve the energy equations [21–23]. However, the generalizability of the solutions was limited by prototype structure, testing approaches, and prediction methods.

CFD is a prevalent tool in evaluating naturally ventilated buildings. Omrani [24] enumerated and assessed the advantages and disadvantages of various methods used to study natural ventilation in multi-storey buildings. In contrast to the time-consuming and cost-intensive experimental methods, once the CFD model has been verified and validated against experimental data, CFD can provide detailed and reliable information on airflow and ventilation performance throughout the space. At the same time, the effectiveness of the commercial software (i.e., Fire Dynamics Simulator, FDS) based on the CFD model in the field of natural ventilation has been verified and validated. It is found that the numerically predicted volume flow rate through the chimney cavity fits reasonably well with experimental data under natural ventilation [25, 26].

Through extensive literature review, most quantitative analyses and parameterization modeling focus on the simple case of a single room and separate SC air channels [27–30]. In a comprehensive review study, Shi et al. [31] grouped thirteen main influencing factors into four categories and compared the corresponding empirical equations with those reported in the literature. Unlike other passive ventilation systems with air layers, the existence of the heat-absorbing wall makes the SC embrace the function of space cooling and ventilation with a function of warm air supply. Li [32] derived airflow rate and thermal stratification solutions in a single-zone building with buoyancy-driven ventilation. The traditional model proposed by Andersen [33] was evaluated. Shi [27] developed four theoretical models to predict natural ventilation under various modes, such as space cooling and heating. Quantitative results obtained from analytical models have been proven acceptable through experimental validations. Although the traditional model is easy to use in predicting ventilation rate, it is at the cost of realistic scene accuracy. To this end, Cheng et al. [34] developed a theoretical model considering the vertically linear and horizontally parabolic distribution of the cavity air temperature. In [35], Tao et al. summarized analytical models for independent vertical SC air channels.

As the parametric studies of single-storey SC continue to improve, researchers focus on establishing a simplified model of solar-induced buoyant ventilation in multi-storey buildings. A CFD model verified by experimental data has addressed the ventilation effects of a two-storey building with a slanted roof SC in Ref. [36]. It was concluded that the qualitative results obtained from the analytical models underestimated the airflow rate compared with CFD simulations. Asadi et al. [37] addressed the effects of SCs installed in four different directions in a seven-storey office building on the indoor ventilation rate with the aid of EnergyPlus. It was found that each SC could offer appropriate ventilation for the neighboring areas.

However, the generalized evidence of SC-induced natural ventilation in multi-storey buildings is still inconclusive. There is a typical structure throughout the air channels used in building envelopes, such as SCs and double-skin façades [38]. It also was found that the atrium enhancement lessens with a higher building due to the atrium's decreased impact on stack pressure and ventilation flow. This is similar to the dilemma when installing SCs in multi-storey buildings [39, 40]. Although buoyancy ventilation systems have some common characteristics, straightforwardly describing the flow transition process and heat transfer with multiple inlets and a single outlet is still challenging. The ventilation effects of multi-storey SC buildings remain elusive and require extensive numerical simulations and experimental case studies.

Simplified mathematical models are most suited for early design phases, when a parametric examination of alternative design configurations may be required to analyze numerous design possibilities [24]. Following the loop approach described by Refs. [41–43], Acred and Hunt [44] developed a simplified mathematical approach to modeling atrium/heated atrium ventilation in multi-compartment buildings with an inlet plenum and open-plan occupied spaces. The flow governing equations of a three-storey building were solved numerically for the case of equal heat inputs on all storeys. However, the flow and thermal behaviors of an atrium and an SC may vary owing to the diverse geometries of various passive solar-assisted systems, leading to a varying distribution of absorbed radiation, temperature, and airflow. Literature on this aspect of multi-storey SC is currently limited, but the underlying buoyancy-driven principles and structural features with air channels are similar, making the study of multi-storey atrium an excellent reference to SCs.

It is noted that a comprehensive assessment that affects the multi-storey SC performance based on design parameters and solar radiation intensity is lacking. Moreover, generally simplified mathematical methods capable of hand-calculating volume flow rate for the case of unequal heat inputs on all storeys have not been fully developed. Four questions related to SC-driven natural ventilation solutions for multi-storey buildings arise, namely: (a) how do core variables affect ventilation capacity and flow patterns in multi-storey buildings; (b) how to get the flow rate of each floor with a small computational overhead; (c) how to adjust the design parameters for maximizing the ventilation rates, and (d) what the differences in applicability between proposed mathematical approaches are.

As a pioneering work of multi-storey buildings with an SC, the specific objective of this study was to expand SC applications from single-zone/floor buildings to multi-storey buildings. Three mathematical methods (including analytical approaches and empirical equations) applicable to different floor quantities were carried out, and numerical simulations with varying parameters of design were developed, addressing the following points:

- Providing an intuitive understanding of physical processes of airflow and heat through SC-induced buoyant ventilation in an example two-storey building through multi-scenario numerical simulations;

- Revealing the relationship between the ventilation performance measured in terms of volume flow rates of multi-storey SC buildings and each core design variable through a simplified mathematical method developed by a single-storey SC theoretical model combined with the closed flow loops method;
- Comparing the applicability and limitations of the three mathematical methods by substituting the design scenario parameters of the experimentally verified numerical model and solving the three mathematical methods; and
- Providing a reference for future designs of wall SCs incorporated into multi-storey buildings.

The rest parts of this paper are organized as follows. Section 6.2 introduces the establishment of numerical models and scenarios assignment. In Sect. 6.3, the ventilation potential of a target building under the influence of design parameters is visualized through an experimentally validated numerical model. In Sect. 6.4, based on the traditional model, we proposed schemes for establishing mathematical models for multi-storey buildings. Then, the example of a two-storey case is compared against numerical simulations for validation. Section 6.5 discusses the scopes and limitations of application for mathematical approaches. Conclusions are presented in Section 6.

6.2 Numerical Procedure

6.2.1 Modeling Basis

The FDS model is utilized for the three-dimensional simulation. The physical domain of an example two-storey building examined is shown in Fig. 6.1. The building's front wall and the SC's glass wall are translucent to illustrate the structure's composition. A continuous and uneven heat flux heats the opaque absorption wall to simulate the radiation absorption and transmittance of the SC, as illustrated by the red area.

The absorption and glazing walls, connected to the building's façade on one side, create the SC's air channel/cavity gap (d). The air inlet height is variable and denoted by h. The vertical distance between the floor and the inlet center (h_{in}) represents the air inlet position. The position of each air inlet can be moved from the floor level to the ceiling level. h_{sc} represents the total height of an SC, and the maximum height in this study is more than twice the height of the roof. w_c stands for the chimney width.

The dimensions of the two rooms of this building are identical, namely 6.0 m (L_r) × 6.0 m (W_r) × 3.0 m (H_r). Orifices positioned in the center of each floor's left wall represent the window of the living space and are always square, connecting directly to the external environment. In the reference case, dimensions include an SC with 0.3 m air channel/cavity gap (d), 0.3 m air inlet height (h), 6.0 m chimney width (w_c), and 6.3 m high chimney cavity (equal to the height of the building). The dimension of the windows is 1.0 m × 1.0 m (A_{open}).

The parameter setting range and numerical scenarios in this study are detailed in Table 6.1. The influencing parameters that affect the ventilation performance of

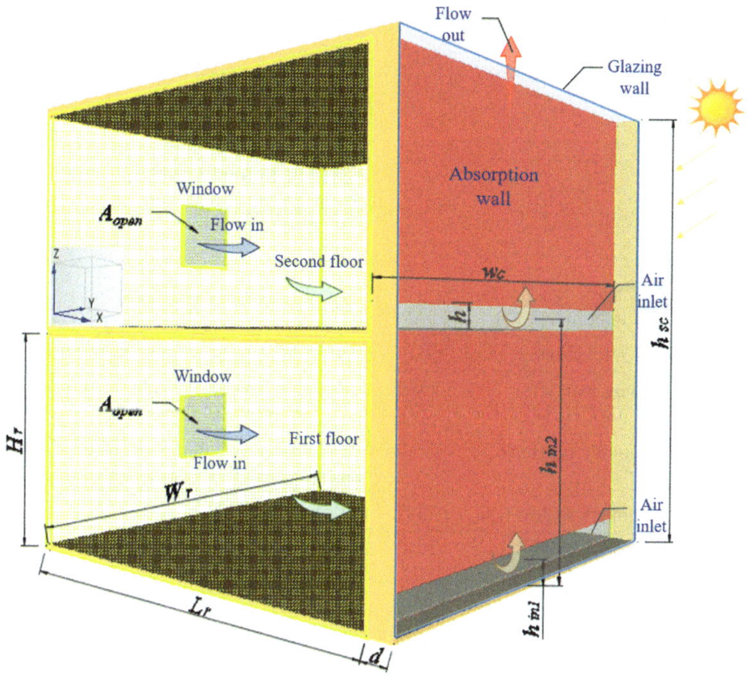

Fig. 6.1 SC and multi-zone numerical model

SCs build on evidence from a prior study [31]. This study evaluated 39 scenarios by adjusting a single design parameter simultaneously, while scenario No. 0 serves as a control/reference case.

6.2.2 Governing Equations

The present study developed the CFD models to examine the ventilation performance of a two-storey SC building. FDS solves the governing Navier–Stokes equations to directly solve for the fluid dynamic properties governing ventilation flow [24]. To discretize partial differential equations, the finite volume method is employed.

Continuity equation can be given by

$$\frac{\partial \rho}{\partial t} + \nabla \cdot \rho \vec{u} = 0 \tag{6.1}$$

Momentum conservation equation is written as

$$\rho\left(\frac{\partial \vec{u}}{\partial t} + (\vec{u} \cdot \nabla)\vec{u}\right) + \nabla p = \rho g + f + \nabla \cdot \vec{\tau} \tag{6.2}$$

Table 6.1 Scenarios were examined under various influencing factors

Influencing factors		No.	Values[a]	Interval	Unit
Control/reference case		0	Two storeys. Each floor with dimensions 6.0 m × 6.0 m × 3.0 m; chimney height: 6.3 m; inlet size: 0.3 m; cavity gap: 0.3 m; inlet location[b]: 0 m; window area: 1.0 m square; solar intensity: 500 W/m^2	–	–
Chimney	Air inlet height (h)	1–4	0.1–0.5	0.1	m
	Air inlet location (h_{in})	5–13	0.15–2.85	0.3	m
	Cavity gap (d)	14–20	0.1, 0.2, 0.5, 0.8, 1.0, 1.2, 1.5	–	m
	Chimney height (h_{sc})	21–24	1.5–6.0 m above the building roof	1.5	m
	Chimney width (w_c)	25–29	1.0–5.0	1.0	m
Room	Window area (A_{open})	30–33	1.2 × 1.2, 1.5 × 1.5, 1.8 × 1.8, 2.0 × 2.0	–	m^2
Environment	Radiation intensity (q)	34–38	200–1000	200	W/m^2

Notes [a]The reference case serves as the control of the other scenarios; and [b]It represents the vertical distance between air inlet center and floor

Energy conservation equation can be described by

$$\frac{\partial}{\partial t}(\rho h) + \nabla \cdot \rho h \vec{u} = \frac{Dp}{Dt} - \nabla \cdot \boldsymbol{q} + \varepsilon \tag{6.3}$$

where t stands for the time, ρ is the air density, $\vec{u}$ is the velocity vector, p denotes the pressure, $\vec{\tau}$ denotes the viscous stress tensor, f represents the external force vector, $\nabla \cdot \boldsymbol{q}$ indicates the conductive and radiative heat fluxes, h denotes the enthalpy, and ε represents the viscosity dissipation rate.

Large eddy simulation (LES) achieves scale separation via the use of a low-pass filter with width Δ—the cube root of the cell volume ($\Delta = V_c^{1/3}$, where $V_c = dxdydz$) [45] features that are greater than the width. A filtered arbitrary continuous field Φ is defined as:

$$\Phi(x, y, z, t) = \frac{1}{V_c} \int_{x-\delta x/2}^{x+\delta x/2} \int_{y-\delta y/2}^{y+\delta y/2} \int_{z-\delta z/2}^{z+\delta z/2} \phi(x', y', z', t) dx' dy' dz' \tag{6.4}$$

Thermal radiation's effect on the energy conservation equation was modeled using Deardorff's model. An algorithm that uses a default grey gas radiation transfer model is expressed as follows [46]:

$$s \cdot \nabla I_\lambda(x, s) = \kappa(x, \lambda)[I_b(x) - I(x, s)] \tag{6.5}$$

where $I_b(x)$ denotes a Planck function source term, $I_\lambda(x, s)$ denotes the radiation intensity at wavelength λ, $\kappa(x, \lambda)$ denotes the absorption coefficient at an x of the wavelength λ, and s denotes the unit normal direction vector.

6.2.3 Boundary Conditions and Grid Independence

FDS was employed to predict the airflow rate driven by given heat fluxes. References have reported its viability in predicting the natural ventilation flow [47]. A time-dependent three-dimensional momentum and fluid continuity model provides more detailed information than the conventional time-average turbulence models. In forecasting airflow rates and particle trajectories, the prediction is consistent with theoretical estimations and the corresponding experimental results [48].

Ambient temperature is the temperature of everything at the start of the simulation. The default pressure ambient temperature and relative humidity values were 101.325 kPa, 20 °C, and 40%, respectively. By default, ambient conditions are assumed to exist beyond the 'OPEN' vent, enabling air to enter or exit based on pressure differentials. A default FDS model describing convective heat transfer between the heated wall and ambient air was used to expedite convergence [49]. All solid building surfaces are inert and smooth except for the heated absorption wall [45]. This study specifies the convective heat transfer coefficient as 15.0 W/(m^2·K) [50], and the glazing wall transmittance was set at 0.76 [25]. A continuous and uniform heat flux heats the absorption wall to simulate radiation absorption, ranging from 200 to 1000 W/m^2. The design of an SC was carried out without considering the wind.

The simulation's grid system consists of approximately 2.4–4.1 million grids in all simulation cases, with high-density grids applied around the air inlet and outlet. The grid independence of the FDS simulations has been investigated. Mesh cell sizes of 0.025 m, 0.05 m, and 0.1 m were examined, referring to refs. [25, 26]. The flow rate at the outlet decreases slightly when the grid size is reduced from 0.05 m to 0.025 m. Accordingly, on the premise of not affecting the calculation accuracy, two mesh cell sizes of 0.1 m and 0.05 m in three directions (i.e., x, y, and z) were then adopted to save calculation time.

The simulation end time was set at 150 s. Our preliminary numerical results showed that the simulation time of 150 s is enough to stabilize the volume flow. Besides, the ventilation performance is represented by the vent airflow rates. The volume flow rates were averaged over the last 30 s (between 120 and 150 s). The time step is adjusted dynamically in solving discrete equations using the FDS solver, thus saving the calculation time. For more details can, refer to Ref. [51].

6.2.4 Validation of Numerical Tool by Experimental Data

This study adopted Fire Dynamic Simulator (FDS) based on LES to model the prototype structure 1:1 numerically. Considering the similarity of the arrangement of rooms and chimneys, a previous experiment was used to verify the feasibility of the numerical tool. In line with the experimental settings, the numerical models were validated against published experimental evidence undertaken by Bouchair [52]. The experiment built a wooden room with dimensions of 1.6 m (L) × 1.8 m (W) × 2.0 m (H) and placed it in a large enough building for testing under a steady state. One side of the room is connected to the wall chimney cavity through a single air inlet with a fixed width of 1.4 m, and the other is connected to the ambient environment through a window. The ambient air temperature was kept at 20 °C. The absorption and glass walls were uniformly heated at four temperatures (30, 40, 50, and 60 °C). An insulation board was applied to the wall between the chimney cavity and the room to prevent heat transfer from the chimney cavity to the room.

Following the experimental scenarios, the numerical model investigated the effect of cavity gap, cavity inlet size, and window size on the airflow rate by varying them. The simulation model was simulated using a default model in FDS that explains convective heat transfer between the heated wall and the surrounding air. Subsequently, the numerical results were compared with corresponding experimental findings under different scenarios, including two air inlet heights, four different surface temperatures, and five air gaps. The comparisons showed that the numerical volume flows through the chimney cavity corresponded well to the experimental data collected during natural ventilation. More details related to the experimental methodology can refer to our previous study [25, 45].

While there was an increasing discrepancy in numerical and experimental data under a wider cavity gap, the numerical results demonstrate consistency with the experimental results. Particularly in scenarios with a cavity gap of 0.1–0.2 m, there was excellent agreement between numerical models and experimental findings. Refs. [25, 26, 34] also demonstrated that FDS could resolve airflow rate trends and overall performance in various configurations. The average absolute error is 21.1%. The turbulent airflow within the room underneath the air inertia from the window may be the source of the substantial discrepancies seen as the air inlet approaches the floor and ceiling levels. Besides, the experimental setup is based on specific test geometry and materials and has inherent limitations. Since the numerical tool can provide an acceptable prediction of SCs, further case studies are taken by FDS.

6.3 Numerical Results and Discussion

The FDS model demonstrates the physical properties of flow regimes and the heat transfer process for a two-storey SC building. The velocity fields and cavity temperature profiles can be observed through visualization. The numerical model in this study operates with a constant flow of fresh air supply and ensures a unidirectional flow. It implies that the indoor air temperature tends to be ambient, and only the convective heat flux in the chimney cavity drives the airflow and produces thermal stratification.

6.3.1 Flow Behavior and Temperature Distribution

Compared to other buoyancy-driven ventilation systems with an air layer, the stack effect of the SC is more robust due to the high absorptance of the opaque absorber wall [35]. The SC mainly absorbs solar radiation through an absorbing wall with insulating properties, so the heat gain in the cavity can be approximately expressed as the radiation absorbed by the system. Solar radiation heats up the absorber wall through non-contact heat transmitted through the outer glass cover, causing a temperature gradient in the vertical air channel, thereby generating thermal buoyancy to promote airflow through the air channel. Each floor has a single air inlet connected to the chimney. The pressure difference caused by the airflow network in the air channel then drives the air movement from the adjacent spaces to the air channel through the flow paths (i.e., vents) to discharge the stagnant indoor air through the outlet.

Figure 6.2a illustrates the velocity-changing tendency of the benchmark case. The distinct flow transport with substantially higher velocities can be observed in the air channel and the central portion of the naturally ventilating zones on each floor. Due to the siphoning effect, the airflow induced in the room's central section through windows can be easily transmitted to the SC's air inlets. Lower floors benefit from a more significant buoyancy effect; thus, they flow faster than higher floors. The initial airspeed at the air inlet further affects the volume flow rates through the room into the air channel.

In the air channel, the velocities promoted by the heat flux near the absorption wall are greater than those in the remaining region. When the cavity air is heated and discharged through the air outlet, the static pressure drops rapidly, inducing the indoor air to flow into the air channel from the air inlet. The near-wall areas of the absorber have higher velocities than the surrounding area because the air near the absorption wall warms up faster. Since the airflow of the first and second floors move upward and superimpose, the cavity airspeed increases with the increasing height along the air channel. Regions with high airspeed also mean a higher temperature difference, referring to the temperature distribution in Fig. 6.2b. The temperature differential in the chimney cavity caused by the heated wall setting produces a difference in air density, and the input heat flux propels the entire process. Heated air continues to flow upward and laterally, resulting in faster airflow rates in the top portion of the cavity's center region and near the glass wall than those in a lower portion. The temperature propagates horizontally into the middle of the air channel of the first floor, and the remaining region's air is unheated.

As the velocity of ascending cavity air increases, the plume trajectory diffuses to the central region of the air channel and the outer glass wall under the action of

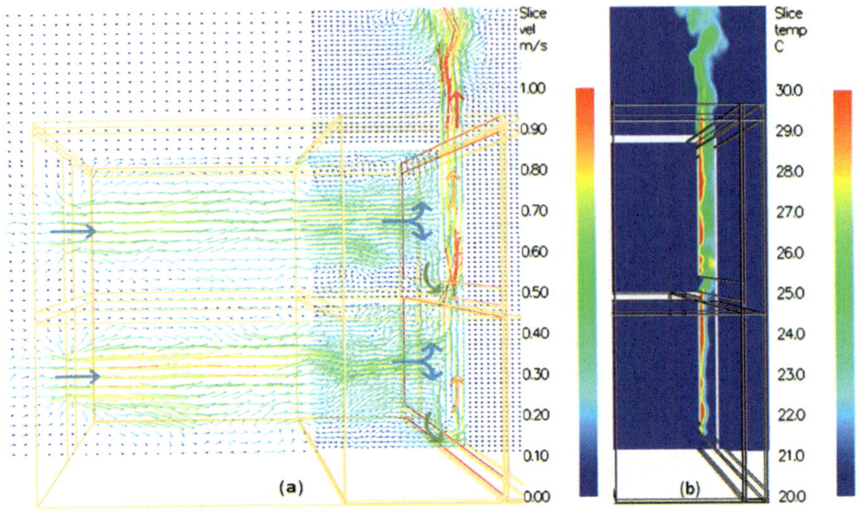

Fig. 6.2 **a** Air velocity vector of side view plot in the building's mid-plane for the reference case and **b** vertical temperature profile in the air cavity at $y = 3.0$ m under 500 W/m^2

pressure difference. Although a buoyancy-driven ventilation system with a vertical air interlayer, such as a double-skin façade (DSF), enjoys some common mechanisms, the existence of the absorption/heated wall makes the fields of velocity, temperature, and flow patterns of an SC different from a single-channel DSF referring to Refs. [38, 53, 54].

6.3.2 Chimney Configurations

The study demonstrates and assesses a correlation between the SC's geometric parameters and volume flow rates in this section. To present the flow trajectory and natural convection, the following also gives the behavioral response of the velocity and temperature field under different factors. When results show a monotonic increasing straight line, the least-square regression curve was fitted on the data set points for acquiring the correlations [5, 31, 35].

Figure 6.3a illustrates how air inlet height (h) influences flow rates. The total outlet flow improves gradually with the air inlet height increase. However, the volume flow rate falls with the inlet height exceeding 0.3 m. The outlet volume flow rate (V_{out}) ranges from 1.03 to 1.24 m^3/s. Analogously, the volume flow rate at the air inlet on the first floor (V_{in1}) is more pronounced to peak at the inlet height between 0.2 and 0.3 m. When the inlet height exceeds the critical value, the volume flow rate falls from 0.68 to 0.58 m^3/s. Differently, the 2nd-floor flow rate at the air inlet (V_{in2}) has progressively grown, ranging from 0.42 to 0.57 m^3/s. Nevertheless, the growth rate becomes moderate when the air inlet's height on the second floor exceeds 0.3 m.

Therefore, changing trends imply that the optimal inlet height is between 0.2 m and 0.3 m, which is consistent with the single-chamber and single-floor SC design [55, 56].

Figure 6.4 shows the influence of different chimney configurations on the cavity air temperature along the vertical air channel. The temperature variation curves along the middle section of the air channel show that the air temperature is bifurcated at half-building height under the continuous convective heat flux. Figure 6.4a shows the

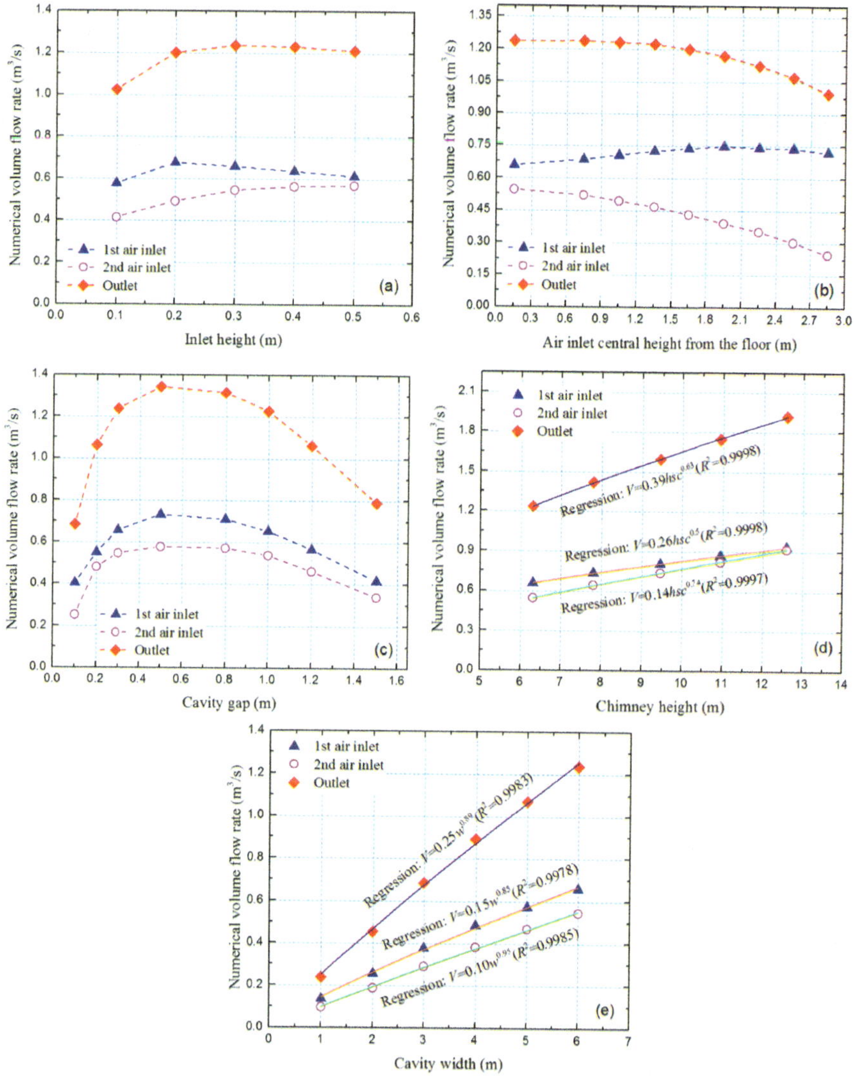

Fig. 6.3 Impacts of chimney configurations on the volume flow rate: **a** inlet height; **b** inlet location; **c** cavity gap; **d** chimney height; and **e** chimney width

cavity temperature changing trajectories with air inlet height increase at 0.15 m from the air inlet. When the air inlet height is 0.1 m, the heated area is the largest, making the temperature at the center of the cavity greater than in other cases in this group. The temperature variation in the middle of the chimney cavity tends to be insignificant whether the air inlet height grows from 0.2 m to 0.5 m. Figure 6.4a, 6.5a show that the longitudinal air temperature distribution in the chimney cavity has experienced a leap from the initial linear increase to the second floor. The cavity air temperature of the second floor is higher than that of the first floor, but the growth trend is decreasing. The more accurate temperature distribution in the chimney described in reference [34] is vertically linear and horizontally parabolic temperature distribution.

Since the value of the surface convective heat transfer coefficient is related to the shape and position of the heat transfer surface, and the flow rate of the working fluid, although the air velocity loss reduces as the air inlet height rises, the decrease of effective heat area weakens the volume flow rates for 1st floor. The greater the flow velocity near the surface, the greater the convective heat transfer coefficient of the surface, then the greater the amount of heat dissipation/absorption. The variation trends of ventilation rates agree with observations of multi-storey SC. The excessive

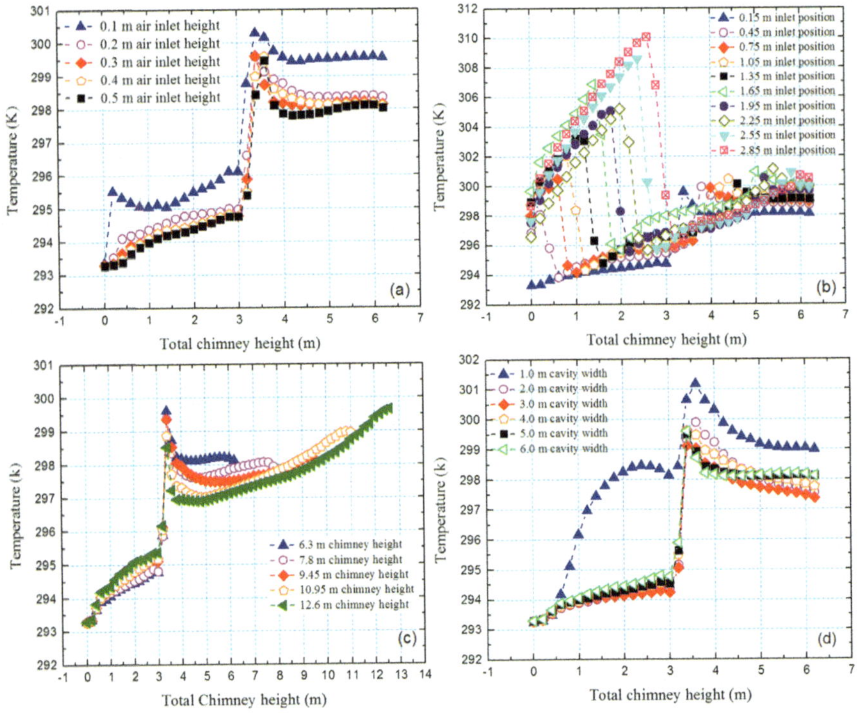

Fig. 6.4 A comparison of the influence of chimney configuration on the distribution of the air temperature in air channels (with a heat flux of 500 W/m^2): **a** inlet height; **b** inlet location, **c** with different chimney height, and **d** at different cavity widths

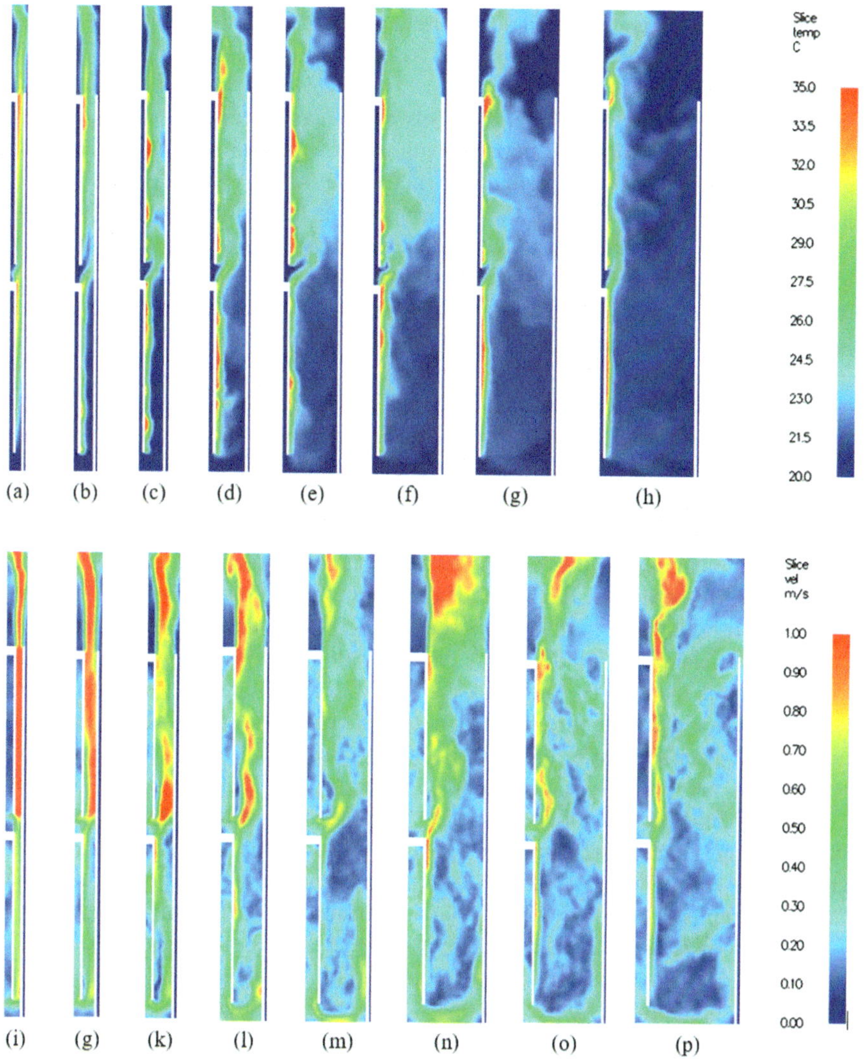

Fig. 6.5 Cavity temperature contour with different gaps (with a chimney height of 6.3 m and heat flux of 500 W/m^2) and corresponding cavity velocity profiles: Temperature: **a** 0.1 m; **b**0.2 m; **c** 0.3 m; **d** 0.5 m; **e** 0.8 m; **f** 1.0 m; **g** 1.2 m, and **h** 1.5 m; Velocity: **i** 0.1 m; **j** 0.2 m;**k** 0.3 m; **l** 0.5 m; **m** 0.8 m; **n** 1.0 m; **o** 1.2 m, and **p** 1.5 m

ventilation flow rate of the 1st floor reduces the air discharge, which is adverse to the accumulation of buoyant air in the air channel [57]. On the other hand, the cavity air temperature increases linearly, and the temperature difference increases the cavity airflow to a certain extent for the 2nd-floor level. Although the flow velocity in the chimney cavity of the 2nd-floor level is higher than that of the 1st-floor level due to the superposition effect, the buoyancy enhancement effect of the chimney at the air inlet on the top floor is weakened due to the shortened height difference between the air inlet and the air outlet. Thereby upper indoor zone always has less flow than the lower floor.

Figure 6.3b indicates the correlation between the air inlet positions (h_{in}) and the volume flow rates. For the first floor, the volume flow rate initially increases, and as the air inlet reaches 1.65 m from the floor (i.e., at the same level as the window), V_{in1} increases from 0.66 to 0.75 m^3/s. As the air inlet rises further, the flow rate gradually drops. However, V_{out} and V_{in2} notice a decreasing tendency when the air inlet shifts upward. Decrease from 1.24 to 0.92 m^3/s, and from 0.55 to 0.17 m^3/s, respectively. The inconsistency is due to the weakening buoyancy effect as the outlet's relative position, and the air inlet is continuously reduced. Besides, the efficient area with heat inputs is lowered as the air inlet upwards.

As shown in Fig. 6.4b, the 1st-floor cavity temperature experienced a rapid improvement before the air inlet location of 1.65 m because the absorption wall under the air inlet continued to heat the cavity air, but then the growth trend slowed down. The upward movement of the air inlet also changes the effective heated area (A_s). The lower part of air is mixed horizontally, increasing the temperature in the lower portion of the cavity and moving upward. However, the temperature of the second floor is lower than that of the 1st floor due to the reduction of the heat-absorbing wall area ($A_{s,2}$) and the interference of the air inlet on the 1st floor. Accordingly, the temperature field variation affects the cavity's velocity distribution. When the first floor's flow rate reaches its maximum, the cavity air in the upper portion exhibits the maximum velocity in its entirety, owing to the superposition effect. The growth trend slows down eventually after the total discharge contributed by the first and second floors at the outlet reaches a maximum.

Figure 6.3c depicts a volume flow rate variation with varied cavity gaps within the parameter range. At a 0.5 m cavity gap, the airflow rate into the air channel from each floor is substantial. The sensitivity of the volume flow rate is remarkable when the cavity gap is lower than 0.5 m. V_{in1} rises from 0.40 to 0.73 m^3/s, V_{in2} is from 0.25 to 0.58 m^3/s, and V_{out} is from 0.69 to 1.34 m^3/s. The turning point of growth tendency occurs when the cavity gap exceeds 0.5 m, and the volume flow rates have ushered in a precipitously drop. It shows that the cavity gap has a negligible impact on ventilation when it exceeds a limit under a constant heat flux imposed on the absorber wall. This is consistent with an optimal value in the cavity gap widening of the monolayer SC, and the larger cavity would relax the flow of kinetic energy [58, 59].

Figure 6.4 shows the FDS predicted temperature contours of the air channels with different cavity gaps. The heat flux's heating effect on the cavity air is concentrated in the adjacent regions of the heat-absorbing wall. Under a heat flux intensity of 500 W/m^2, the heat plume in the 1st-floor height range remains permanently attached to the absorption wall, while the heat plume in the chimney cavity of the 2nd floor spreads to the outer glass wall. The cavity temperature is higher than 23 °C with a gap of 0.1 m, indicating the cavity air is thoroughly heated due to the given heat flux. Although the temperature of the plume on the 2nd floor spreads laterally and progressively declines as the air channel's breadth rises, the unheated part occupies most of the cavity, and the heat flow is reabsorbed into the absorption wall when the gap exceeds 1.2 m. The overall heating effect on the airflow diminishes as the gap exceeds 0.5 m.

Besides, the cavity gap affects the magnitude of the vertical air channel's limiting its influences on natural convection [38]. Figure 6.5 also reveals that the high-speed zone of the cavity airflow is mainly found along the absorbing walls. When the heat flux and chimney height are constant, the residual portion's air velocity drops and becomes more chaotic as the cavity gap increases. Although increasing the cavity gap initially helps increase the flowrate through the air channel, the improvement becomes insignificant. The heating effectiveness is curbed when the gap reaches a certain value.

The chimney height considerably influences the volume flow rates in both single-storey and multi-storey building cases, substantially affecting the convection evolution by affecting the overall energy input. Figure 6.3d presents that the flow rates improve significantly under a heat flux of 500 W/m^2 and a gap of 0.3 m as the chimney height grows from 6.3 m to 12.6 m. The regression results tie nicely with the prior studies of single-zone cases wherein Ref. [31] ($V \propto H^{2/3}$): $V = 0.26{h_{sc}}^{0.5}$ ($R^2 =$ 0.9998) for the first floor; $V = 0.14{h_{sc}}^{0.74}$ ($R^2 = 0.9997$) for the second floor; $V =$ $0.39{h_{sc}}^{0.63}$ ($R^2 = 0.9998$) for the outlet, where V denotes volume flow rate, and h_{sc} denotes chimney's height. For scenario No. 34 (i.e., the chimney height exceeds the total building height by twice), the flow rate at each vent reaches its peak value, V_{out} increases from 1.24 to 1.92 m^3/s, V_{in1} increases from 0.66 to 0.94 m^3/s, and V_{in2} improves from 0.55 to 0.92 m^3/s. Figure 6.4c shows that the central region's changing tendencies of cavity temperature is almost identical as the chimney height grows. The effects of chimney width on volume rates are similar to those of chimney height. The adjustable chimney width changes the outlet area and the absorption wall's heating area. Reducing the breadth of the air channel unnecessarily is detrimental to boosting airflow rates.

6.3.3 Window Area

The volume flow rates rise marginally with an increasing window area. The regression curves, as seen in Fig. 6.6, show approximate fit power growth trends $V = 0.68A_{open}^{0.17}$ ($R^2 = 0.9602$) for the first floor; $V = 0.56A_{open}^{0.05}$ ($R^2 = 0.6820$) for the second floor; and $V = 1.27A_{open}^{0.12}$ ($R^2 = 0.9305$) for the outlet, where A_{open} denotes the window area. The volume flow rates of the first floor and the outlet exhibit similar progressive rising trends as the window area rises from 1 to 4 m^2, ranging from 0.66 to 0.85 m^3/s for the first floor and 1.24 to 1.47 m^3/s for the outlet, respectively. However, the volume flow rate of the second floor improves slightly with the window area increase, ranging from 0.55 to 0.59 m^3/s. In the ventilation process driven entirely by buoyancy, the air induced to flow due to the pressure difference is supplemented by fresh air through the windows. As the window area is over $1.5 \times 1.5\ m^2$, the growth rates

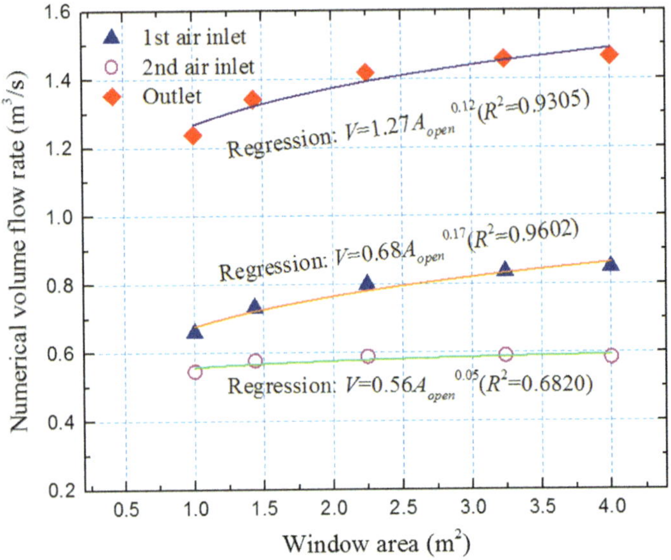

Fig. 6.6 Effect of window area on volume flow rates

of volume flow rates decline. When the window area enlarges, it is imperceptible to detect the flow rate change at the air inlet on the second floor.

The present study provides additional evidence concerning the effect of window area on flow rates. The dependent manner of the airflow rate of a single chamber on the window size shows a similar expression in Ref. [45]. When comparing the empirical equation: $V_{\text{in}} = A_{\text{open}}/\left(0.65 + 5.2\, A_{\text{open}}\right)$ by inductive reasoning to the prior research, it can be seen that the window openings of the building do not always favor ventilation efficiency. The impact of window area on the growth rate of volume flow rates is not evident due to the velocity loss of airflow during transmission in the rooms and the flow restriction imposed by the flow of the room sidewalls.

6.3.4 External Environmental Conditions

Solar-assisted passive ventilation is heavily reliant on solar radiation intensity. The volume flow of the three locations thus exhibits a power function growth with the radiation intensity varying from 200 to 1000 W/m^2, as illustrated in Fig. 6.7. The relative area of the heat-absorbing wall on the first floor is more significant than on the second floor. So, under the same solar radiation intensity, the solar energy received by the first floor is greater than that of the second floor, resulting in a higher volume flow rate.

The relationship between flow rates and heat flux intensity can be approximated as power relationships: $V \propto q^{1/3}$, where q denotes the solar radiation intensity, W/

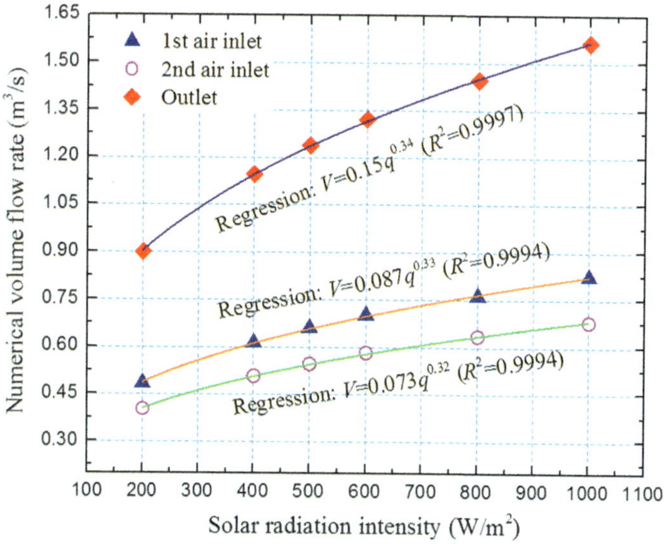

Fig. 6.7 Effect of solar radiation intensity on airflow rates

m². This variation trend is similar to those results from simple cases of single-room research. A similar exponential correlation $V \propto q^{0.35-0.76}$ was reported in single-room cases [60–62]. From the exponent in the power function, it is found that as the volume and amount of the ventilated space increase, the influence of solar radiant energy on the volume flow rate decreases slightly.

Numerical findings provide some instantly reliable inferences. There are specific values to maximize the ventilation rate. One possible reason to account for these results is that initial air velocity at the air inlet, transmission distance, and heat release rate determine the flow trajectory and heat exchange efficiency. The visualized trajectory of the velocity and cavity temperature fields provides an intuitive explanation of the change in volume flow rates. The simulation results also show that for all scenarios in this study, the average range of the cavity temperature is 21.7–28.6 °C under 200–1000 W/m².

The changing tendencies of the volume flow rates with different influence factors discussed above are summarized. The strength of the buoyancy effect has a decisive influence on the ventilation capacity. Namely, the main design parameters related to a straight enhancement in buoyancy, including effective heated area (A_s) and net heat flux (i.e., solar radiation intensity q), have the most significant impact on volume flow rates. The controlling factors that can significantly change the magnitude of the above design parameters include chimney height and width, the relative distance between the air inlet and the air outlet, and solar radiation intensity. They show similar influence trends since these factors govern the energy input into the SC's flow and heat transfer procedure.

Design parameters (i.e., cavity gap and air inlet size) that influence the cavity's temperature and velocity distribution and the initial air velocity at the inlet produce inflection points in the volume flow trajectory. They determine the extent of the restricting effect on the natural convection. The air flow rates initially improve with these parameters enlarging, and then there is a downward trend. Therefore, when the heat flux and chimney height are constant, if the cavity gap and inlet height exceed 0.5 m and 0.3 m, respectively, the effect on the flowrate becomes insignificant.

6.4 Mathematical Models

So far, solar radiation and design parameters that affect SC performance incorporated into an example two-storey building have been discussed numerically. This section will discuss the establishment of simplified mathematical models using the core variables identified by the numerical simulations above. The main goal of the current study is to develop easy-to-apply mathematical methods to advance a deeper insight into the practice of solar-chimney-based buoyancy ventilation in multi-storey buildings and the current challenges. Three mathematical methods of the simplified model are proposed based on a traditional model that the airflow in the chimney cavity is regarded to be well-mixed and uniform [27, 34] under given conditions. The ventilation rates of the three methods are compared with corresponding numerical findings. The room area is a typical one on both floors, as the impacts of the floor area on the overall ventilation performance can be disregarded in light of our prior findings [27].

The factors influencing SC performance and the interaction between airflows from air inlets on different floors complicate the processes. Uncertainty exists due to the increasing number of openings, the interaction of flow rates on each floor, and the difficulties in resolving nonlinear solution multiplicity for naturally ventilated multi-storey buildings. Potential techniques to solve problems in theoretical developments are using air bifurcation phenomena [63, 64] and the closed flow loop method [41] to expand SC applications from single-zone/floor buildings. Namely, when a single chimney cavity is connected with multiple air inlets and a single outlet, there is a relationship between the total discharged flow rate and each floor flow rate for distribution, as illustrated in Fig. 6.8.

6.4.1 *Physical and Mathematical Models*

A two-dimensional physical model for natural ventilation is established. Geometric parameter nomenclature is exhibited in Fig. 6.1 and Table 6.1. Pressure losses are considered across windows, air inlets, and an outlet. Table 6.2 briefly introduces the three mathematical methods.

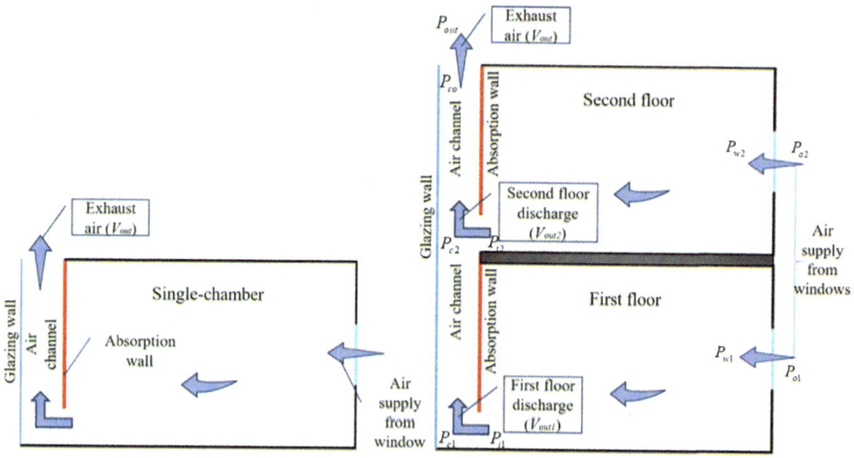

Fig. 6.8 SC in **a** a single-storey building; and **b** a multi-storey building showing air bifurcation. *Note* P denotes pressure losses across floor windows ($A_{w,i}$), inlets ($A_{\text{in},i}$), and air outlet (A_{out})

Table 6.2 Three methods and their explanation

Methods	Explanation
Method 1	The flow rate contributed by each floor is proportional to the overall discharge flow rate, and the analytical solution such a proportional factor can be obtained;
Method 2	Solve a set of nonlinear equations to obtain analytical solutions for volumetric flow rates;
Method 3	Calculate the flow rate of each floor independently according to a single-zone analytical model and combine the results of numerical modeling to obtain the empirical equation

For the control volume, equations for mass balance, energy conservation, and momentum can be obtained. The relationship between inlet pressure difference and flow velocity through an orifice can be given by:

$$\Delta P_i = \left(\frac{V_i}{C_d A_i}\right)^2 \frac{\rho_i}{2} \tag{6.6}$$

where ρ denotes air density, kg/m^3; ΔP denotes the pressure difference, Pa; C_d represents the discharge coefficient; V stands for the volume flow rate via the aperture, m^3/s; A denotes the aperture area, m^2; and i represents the aperture number.

Assuming that the air in rooms and the chimney cavity are evenly mixed, aerostatic pressure distribution in different spaces can be expressed as:

$$\Delta P_i = \rho_i g \Delta h \tag{6.7}$$

where Δh denotes vertical altitude difference, *m*; and g denotes the gravitational acceleration, m/s^2.

Fulfillment of mass balance under steady-state conditions leads to the following equation:

$$\rho_0 V_{\text{in},i} = \rho_0 V_{w,i} \tag{6.8}$$

where subscript in represents the air inlet; subscript w stands for the window, and subscript i denotes the number of openings.

The relationship between P_{o1} and P_{out} (also P_{o2} and P_{out}), as shown in Fig. 6.8, can be described by Eq. (6.6). The connection between the pair pressures of each aperture, such as P_{o1} and P_{w1}, P_{i1} and P_{c1}, is given in Eq. (6.7), which also applies to the pressure gradient between both sides of the aperture. Based on the principle of energy conservation, the following equations can be obtained,

$$\rho_c V_{\text{out}}^2 \frac{1}{C_d^2 A_{\text{out}}^2} = 2g(\rho_0 - \rho_c)\left(h_{\text{sc}} - h_{\text{in},i}\right) - \rho_0 V_{\text{in},i}^2 \left(\frac{1}{C_d^2 A_{w,i}^2} + \frac{1}{C_d^2 A_{\text{in},i}^2}\right) \tag{6.9}$$

where ρ_c stands for the average air density within the SC cavity.

Under steady-state circumstances, the energy conservation equation for a chimney cavity is stated as:

$$E_c = q\tau A_{s,i} \tag{6.10}$$

where q denotes the heat flux intensity; E_c identifies the heating intensity power in the chimney cavity, W; τ denotes the glazing wall transmissivity; and $A_{s,i}$ denotes the heating area of each floor.

This study uses an ideal gas law equation to describe the relationship between air density and temperature [27]

$$\rho_0 T_0 = \rho_c T_c \tag{6.11}$$

So the thermal heat gain (absorbed heat) is calculated as [27, 65]

$$E_c = \rho_c C_p V_{\text{out},i}(T_c - T_0) \tag{6.12}$$

The constants used in the mathematical method to describe the physical process are shown in Table 6.3.

Table 6.3 Modeling inputs of the theoretical models [27]

Constants	Nomenclature	Value	Unit
C_d	Discharge coefficient	0.6	–
C_p	Specific heat capacity of air	1005	J/kg · K
ρ_0	Density of ambient air	1.205	kg/m^3
g	Gravity acceleration	9.81	m/s^2
T_0	Ambient temperature	293.15	K
τ	Glass transmissivity	0.76	–

6.4.2 Method 1

Bifurcation phenomena are assumed to occur in the flow field, and the proportional coefficient can characterize the constant fluid properties. Considering that the discharged airflows from two rooms are connected through the chimney cavity, the mass conservation equation can be expressed as

$$\rho_c V_{\text{out}} = \rho_0 (V_{\text{in}1} + V_{\text{in}2}) \tag{6.13}$$

Subsequently, Method 1 introduces unknown variables α_1 and α_2 to represent the concept of airflow bifurcation. Additional independent equations can be expressed by

$$V_{\text{in}1} = \alpha_1 V_{\text{out}} \tag{6.14}$$

$$V_{\text{in}2} = \alpha_2 V_{\text{out}} \tag{6.15}$$

$$\alpha_1 + \alpha_2 = 1 \tag{6.16}$$

Through combining Eqs. (6.9), (6.14), and (6.15), the following governing equation can be obtained,

$$\rho_c V_{\text{out}}^2 \frac{1}{C_d^2 A_{\text{out}}^2} = 2g(\rho_0 - \rho_c)\left(h_{\text{sc}} - h_{\text{in},i}\right) - \rho_0 \alpha_i^2 V_{\text{out}}^2 \left(\frac{1}{C_d^2 A_{w,i}^2} + \frac{1}{C_d^2 A_{\text{in},i}^2}\right) \tag{6.17}$$

To solve simultaneous Eqs. (6.13)–(6.17) and assume $\rho_c/\rho_0 = 1$, the analytical solutions of α_1 and α_2 can be obtained directly. A quadratic equation of one variable can be obtained after elimination.

$$\begin{aligned}(h_{\text{sc}} - h_{\text{in}1}) \times \left[\alpha_1^2\left(\frac{1}{A_{w1}^2} + \frac{1}{A_{\text{in}1}^2}\right) - (1-\alpha_1)^2\left(\frac{1}{A_{w2}^2} + \frac{1}{A_{\text{in}2}^2}\right)\right] \\ -(h_{\text{in}2} - h_{\text{in}1}) \times \left[\frac{1}{A_{\text{out}}^2} + \alpha_1^2\left(\frac{1}{A_{w1}^2} + \frac{1}{A_{\text{in}1}^2}\right)\right] = 0\end{aligned} \tag{6.18}$$

When the solution of α_1 is obtained, V_{out} can be solved by the following equation

$$V_{out} = \sqrt{\frac{2gC_d^2\frac{(\rho_0-\rho_c)}{\rho_0}(h_{sc}-h_{in1})}{\frac{\rho_c}{\rho_0}\frac{1}{A_{out}^2}+\alpha_1^2\left(\frac{1}{A_{w1}^2}+\frac{1}{A_{in1}^2}\right)}} \tag{6.19}$$

Substituting the Eqs. (6.10)–(6.12) into Eq. (6.19). To simplify Eq. (6.19), two coefficients B_i and A_r are introduced and set as follows:

$$B_i = \frac{A_{s,i}q\tau g}{T_0\rho_0C_p} \tag{6.20}$$

where C_p represents the air's specific heat capacity, J/kg · K

$$A_r = \sqrt{\frac{2A_{w1}^2A_{in1}^2A_{out}^2}{A_{w1}^2A_{in1}^2+\alpha_1^2A_{out}^2A_{in1}^2+\alpha_1^2A_{out}^2A_{w1}^2}} \tag{6.21}$$

Rearrange Eq. (6.19) with Eqs. (6.20) and (6.21) to obtain the following equation,

$$V_{out} = (C_dA_r)^{2/3}\times[B_1(h_{sc}-h_{in1})]^{1/3} \tag{6.22}$$

After V_{out} is obtained, V_{in1} and V_{in2} can be obtained by Eqs. (6.14)-(6.16). In the process of using Method 1 to solve α_1, it is noted that due to assumptions and approximations $\rho_c/\rho_0 = 1$, the items in simplified Eq. (6.18) are only related to geometric parameters. This problem has prevented the study from carrying out the flow rates considering solar radiation changes. Besides, there is no analytical solution for scenarios 13–15 (i.e., h_{in1} is 2.55 m and 2.85 m; and the air gap is 0.1 m).

6.4.3 Method 2

This method aims to directly demonstrate the feasibility of solving analytical solutions of volume flow rates through simplified nonlinear equations. An in-house MATLAB code is developed to solve the analytical solutions of V_{in1}, V_{in2}, and V_{out} directly based on Eqs. (6.9) and (6.13). It can be seen from the equations that it is needed to obtain the value of ρ_c, indirectly solving the volume flow rates. According to the numerical results, the temperature in the chimney cavity is maintained at 23–25 °C in various settings, with a radiation intensity of 500 W/m^2. It is assumed that the cavity air density (ρ_c) is a constant in the calculation process of Method 2, that is, the air density value at 25 °C. The accurate estimation of ρ_c stands in the way of obtaining excellent results.

However, there is no analytical solution for scenarios 13–15 (i.e., h_{in1} is 2.55 m and 2.85 m; air gap (d) is 0.1 m) due to the highly nonlinear function of the obtained equations. Besides, this method does not support the consideration of solar radiation since the simplified equation does not contain a temperature term, and scenarios 34–38 (i.e., solar radiation variables) are not calculated.

6.4.4 Method 3

Method 3, as shown in Fig. 6.9, assumes that the airflow discharged from each room is not connected through a chimney but is conserved separately. It is assumed that there is a linear relationship between the total flow discharged from a multi-storey building and the flow discharged from each independent room. An empirical relationship can be expressed as

$$V_{\text{out}} = \alpha_1 V_{\text{out1}} + \alpha_2 V_{\text{out2}} \tag{6.23}$$

where V_{out} represents the total fluid exhaust volume flow rate when two floors share an air channel; V_{out1} represents the calculated volume flow rate of the fluid discharged when the first floor is separated in this prototype building; V_{out2} represents the discharge volume flow calculated separately from the second floor of the prototype building; and α_1 and α_2 denote coefficient term of a linear relationship.

Each room's mass conservation equation is as follows:

$$\rho_c V_{\text{out},i} = \rho_0 V_{\text{in},i} \tag{6.24}$$

where $V_{\text{out},i}$ denotes the discharge flow rate of each floor.

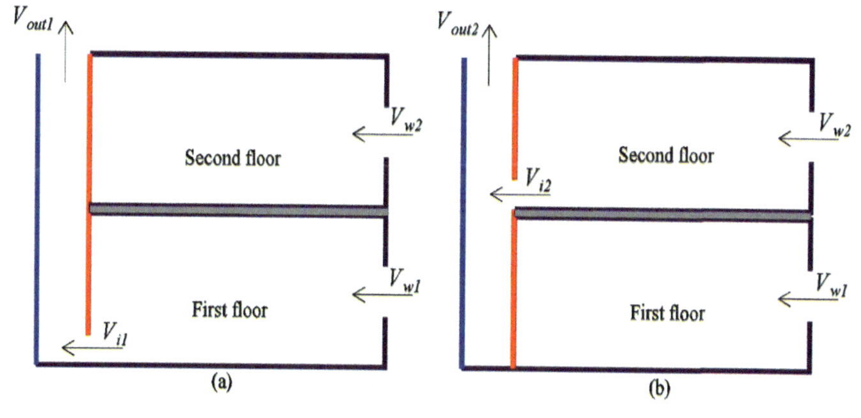

Fig. 6.9 Schematic of Method 3 computational ventilation model for: **a** the 1st floor; and **b** the 2nd floor

According to the calculation process shown in Fig. 6.10, the flow rate of each orifice is considered separately. Equation (6.9) changes as follows:

$$\rho_c V_{\text{out},i}^2 \frac{1}{C_d^2 A_{\text{out}}^2} = 2g(\rho_0 - \rho_c)\left(h_{\text{sc}} - h_{\text{in},i}\right) - \rho_0 V_{\text{in},i}^2 \left(\frac{1}{C_d^2 A_{w,i}^2} + \frac{1}{C_d^2 A_{\text{in},i}^2}\right) \tag{6.25}$$

To simplify the equation, the terms in the equation are arranged according to the characteristics as follows

$$A_i^* = \sqrt{\frac{2A_{w,i}^2 A_{\text{in},i}^2 A_{\text{out}}^2}{A_{\text{in},i}^2 A_{\text{out}}^2 + A_{w,i}^2 A_{\text{out}}^2 + A_{w,i}^2 A_{\text{in},i}^2}} \tag{6.26}$$

Combining the Eqs. (6.20), (6.24)–(6.26), the following equation can be obtained,

$$V_{\text{out},i} = \left(C_d A_i^*\right)^{2/3} \times \left[B_i\left(h_{\text{sc}} - h_{\text{in},i}\right)\right]^{1/3} \tag{6.27}$$

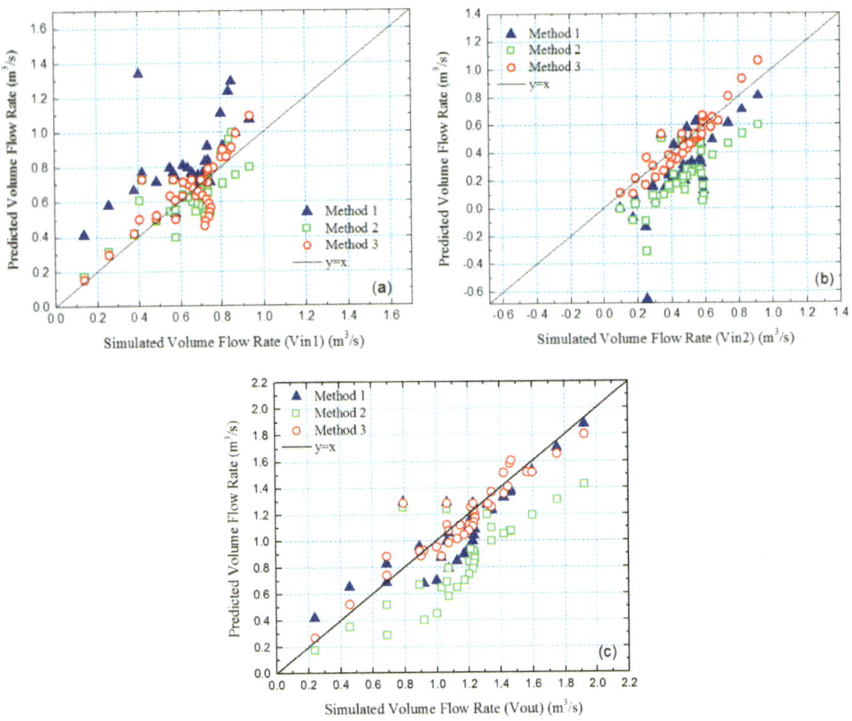

Fig. 6.10 Comparison of computational models via three methods. The negative values in **b** represent no analytical solutions for the three cases based on the in-house MATLAB code that the system offers negative solutions

An in-house MATLAB code was developed to solve the above equations. V_{out} is supplied from numerical simulation. V_{out1} and V_{out2} are obtained by Eq. (6.27). α_1 and α_2 are obtained by linear regression estimation, and the relationship between V_{out}, V_{out1}, and V_{out2} can be obtained by,

$$V_{out} = 1.41V_{out1} - 0.37V_{out2} \tag{6.28}$$

where V_{in1} and V_{in2} can be solved based on the empirical relationship:

$$V_{in1} = 0.67 \times V_{out,1} \tag{6.29}$$

$$V_{in2} = 0.79 \times V_{out,2} \tag{6.30}$$

6.4.5 *Comparisons of the Three Methods*

Figure 6.11 in the appendix presents three mathematical approaches to solve SC-induced buoyant ventilation rates in multi-storey buildings. For calculating constant variables (α) and flow equations, the data inputs were used in conjunction with an in-house MATLAB code and OringinPro.

The applicability of the three mathematical methods is compared with CFD numerical models. Figure 6.10 compares the three mathematical methods for volume flow rates and the volume flow values obtained by CFD numerical models. The discrepancies between the three methods could be due to the simplifications in the postulated conditions.

6.5 Discussion of Mathematical Methods

Regarding Method 1, the analytical solutions are compared with the results of numerical simulations. There are three scenarios without analytical solutions mentioned in Sect. 4.3, and the scenario considers solar radiation variables that are not involved. The average absolute value error between the mathematical method and the numerical simulation of V_{out} is 15.71%. Seventeen out of the thirty-one scenarios are less than 10%. For V_{in1}, the average absolute value error reaches 32.47%. The absolute error of eleven out of thirty-one is less than 10%. For V_{in2}, the absolute value error is 41.48%, and one out of thirty-one scenarios is less than 10%. One possible explanation for the error is that there are potential biases in estimating air density and the air velocity change caused by the temperature profiles.

As for Method 2, after removing the three scenarios without analytical solutions and the scenario considering solar radiation variables, the analytical solutions of the remaining scenarios are compared with the numerical predictions. For V_{out}, the average absolute value error is 28.12%, and two out of the thirty-one scenarios are less than 10%. For V_{in1}, the averaged absolute value error is much lower than the other two methods, reaching 16.35%, and eleven out of the thirty-one is less than 10%. Conversely, for V_{in2}, the averaged absolute value error is 53.63%, much higher than the other two methods, and two out of thirty-one scenarios have an absolute error of less than 10%.

Method 3 has an excellent predictive performance on the air outlet's total air discharge volume flow rate. For V_{out}, the average absolute error is 8.40%, and thirty-one of the thirty-nine scenarios are less than 10%. For V_{in1}, the average absolute value error reaches 14.34%, and eighteen out of the thirty-nine scenarios are less than 10%. For V_{in2}, the average absolute value error reaches 13.74%, and twenty out of thirty-nine scenarios are less than 10%. It is, therefore, likely that such connections exist between each flow. However, the findings of V_{in1} and V_{in2} are mere approximations, and empirical results have supported the hypothesis that there is a linear relationship.

Table 6.4 presents the averaged absolute error between the volumetric flow calculated by the three methods and the numerical simulation result. Method 3 shows a minor error when calculating volume flow rates. Although the prediction results of Method 2 are closer to the numerical simulation when obtaining V_{in1}, it is not ideal for predicting V_{out} and V_{in2}. The prediction error of Method 1 turned out to be a compromise. The introduced proportional coefficients α_1 and α_2 more intuitively reflect the contribution of the discharge flow of each floor to the total outflow.

In the process of simplifying the mathematical calculation to reduce the dimensionality of the equation, various degrees of error are caused. Method 3 develops empirical solutions for predicting volume flow rates, and the results' error is relatively small. Besides, Method 1 can obtain the analytical solution for the number of floors below two and conforms to the intention of the airflow bifurcation hypothesis.

With the similarity and discrepancies in wall SC buildings, the analytical models can be summarized and compared as follows (see Table 6.5), consisting of SC mathematical models reviewed from literature and mathematical approaches developed in this study.

6.6 Conclusions

This study explores the flow characteristics of solar-chimney-based buoyancy ventilation in multi-storey buildings. Research explores evidence to suggest that strategies for improving flow rates involve the optimal chimney geometric parameters, the orifices area on the building envelope, and the radiation intensity. Taking a two-storey prototype building, for example, the ventilation capacity of the SC is analyzed theoretically and numerically. The insights gained from this study can assist in providing design strategies:

Table 6.4 A summary of comparison of developed theoretical models

Methods	Mathematical models	Assumption	Averaged absolute error (%)	Applicability
1	$V_{out} = (C_d A_r)^{2/3}[B_1(h_{sc} - h_{in1})]^{1/3}$	$V_{i1} = \alpha V_{out}$; $V_{i2} = (1-\alpha)V_{out}$	V_{out}: 15.71% V_{in1}: 32.47% V_{in2}: 41.48%	• Pros: air bifurcation hypothesis represented by proportional coefficient is practical; Cons: fails to introduce the solar intensity item
2	$\rho_c V_{out}^2 \frac{1}{C_d^2 A_{out}^2} = 2g(\rho_0 - \rho_c)(h_{sc} - h_{in,i}) - \rho_0 V_{in,i}^2 \left(\frac{1}{C_d^2 A_{w,i}^2} + \frac{1}{C_d^2 A_{in,i}^2}\right)$	$\rho_c V_{out} = \rho_0(V_{in1} + V_{in2})$;	V_{out}: 28.12% V_{in1}: 16.35% V_{in2}: 53.63%	• Pros: direct calculation of nonlinear equations to obtain volume flow rates; • Cons: errors caused by estimated ρ_c; fails to introduce the solar intensity item
3	$V_{out,i} = (C_d A_i^*)^{2/3} \times [B_i(h_{sc} - h_{in,i})]^{1/3}$	$V_{out} = \alpha_1 V_{out1} + \alpha_2 V_{out2}$; $V_{in1} = 0.67 \times V_{out,1}$; $V_{in2} = 0.79 \times V_{out,2}$;	V_{out}: 8.40% V_{in1}: 14.34% V_{in2}: 13.74%	• Pros: easy-to-applied linear relationship; • Cons: empirical evidence not supported by theory

- Optimal geometric factors of the SC exist, leading to the volume flow rate at each orifice reaching a peak. The growth trend slows down or decreases after exceeding the specific design value. The findings illustrate that a wider cavity gap and a higher stack height than the single-zone/floor cases were the most effective ways to enhance the ventilation effect. The air inlet size that produces the maximum flow rates is the same as the single zone/floor;
- The larger window area does not significantly improve the flow rates, especially since the volume flow rate on the second floor is hardly affected by the window area. As the window area exceeds a specific value, the growth rates of volume flow rates decline.

Table 6.5 A summary of analytical models for SC buildings

Case	Mode	Ventilation rate	Assumptions	Solution
Single-room SC buildings	Traditional model	$V_{\mathrm{out}} = \left(C_d A_r^*\right)^{2/3}(\mathrm{B}h_{\mathrm{sc}})^{1/3}$ [27]	Uniform temperature	Analytical
	Advanced traditional model	$V_{\mathrm{in}} = 0.63\left(C_d A_r^*\right)^{2/3}(\mathrm{B}h_{\mathrm{sc}})^{1/3}$ [34]	Vertically linear and horizontal parabolic temperature	Analytical
Multi-storey SC buildings	Traditional model	Method 1 refers to Table 4	Equal storey vent areas, unequal heat inputs	Analytical for two-storey
		Method 2 refers to Table 4		Analytical for two-storey
		Method 3 refers to Table 4		Empirical for buildings with more than three storeys

- The relationship between solar radiation intensity and volume flow rate conforms to a power function $V \propto q^{1/3}$. The radiation intensity is significantly and positively correlated with the flow rates, but it does not contribute as significantly to the SC's dimensions; and
- Method 3 treats two inlets independently and attempts to develop an empirical expression between the single zone and the overall outflow. It can solve the case where an analytical solution cannot be obtained for buildings with more than two floors; Method 1 is suitable for buildings with less than two floors and can obtain analytical solutions to the predicted results through the bifurcation hypothesis.

Further work should shed more light on underlying factors, including dependent variable settings and additional experimental data.

Appendix

See Fig. 6.11.

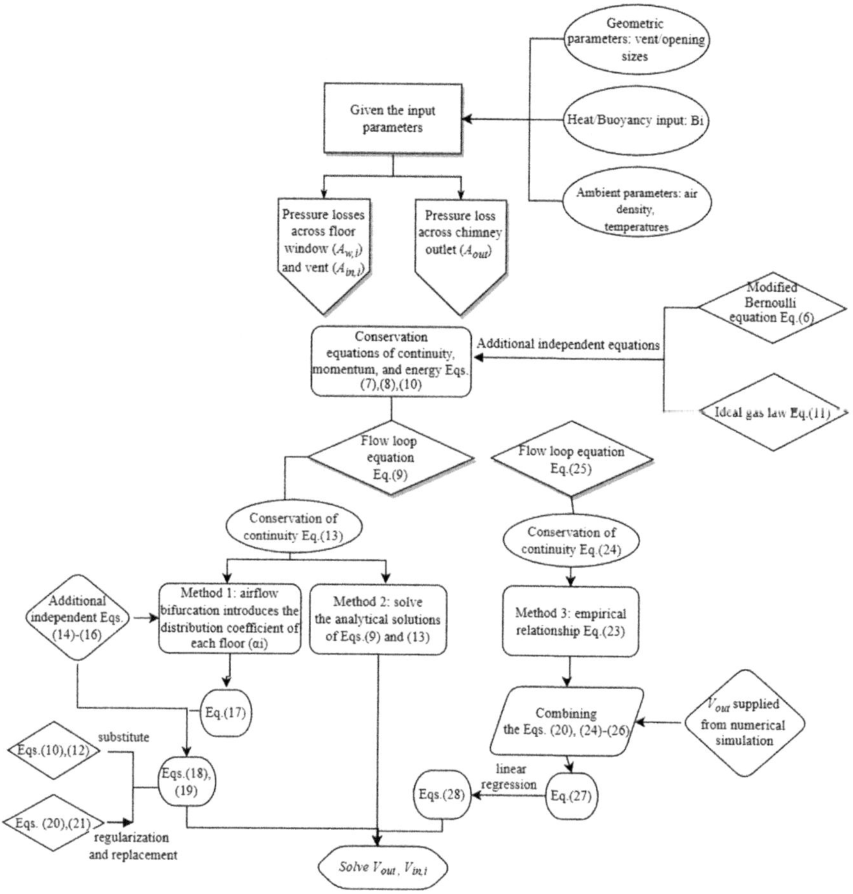

Fig. 6.11 Flowchart of the mathematical approaches in this study

References

1. Alencastro J, Fuertes A, de Wilde P (2018) The relationship between quality defects and the thermal performance of buildings. Renew Sustain Energy Rev 81:883–894
2. Jathar LD et al (2021) Effect of various factors and diverse approaches to enhance the performance of solar stills: a comprehensive review. J Therm Anal Calorimetry 1–32
3. Eli L et al (2021) Thermal performance of residential building with mixed-mode and passive cooling strategies: the Brazilian context. Energy and Buildings 244:111047
4. Esmaili MM et al (2022) Investigation on the performance of a solar chimney-flare gas hybrid system. Sustain Energy Technol Assess 52:102279
5. Goshayeshi HR, Safaei MR (2019) Effect of absorber plate surface shape and glass cover inclination angle on the performance of a passive solar still. Int J Numer Methods Heat Fluid Flow
6. Pourghorban A, Asoodeh H (2022) The impacts of advanced glazing units on annual performance of the Trombe wall systems in cold climates. Sustain Energy Technol Assess 51:101983

7. Sarafraz M, Safaei MR (2019) Diurnal thermal evaluation of an evacuated tube solar collector (ETSC) charged with graphene nanoplatelets-methanol nano-suspension. Renew Energy 142:364–372
8. Sarafraz MM et al (2019) Thermal evaluation of graphene nanoplatelets nanofluid in a fast-responding HP with the potential use in solar systems in smart cities. Appl Sci 9(10):2101
9. Sarafraz M et al (2019) Potential of solar collectors for clean thermal energy production in smart cities using nanofluids: experimental assessment and efficiency improvement. Appl Sci 9(9):1877
10. Safaei MR, Goshayeshi HR, Chaer I (2019) Solar still efficiency enhancement by using graphene oxide/paraffin nano-PCM. Energies 12(10):2002
11. Tyagi V et al (2021) Phase change material based advance solar thermal energy storage systems for building heating and cooling applications: a prospective research approach. Sustain Energy Technol Assess 47:101318
12. Samykano M (2022) Role of phase change materials in thermal energy storage: potential, recent progress and technical challenges. Sustain Energy Technol Assess 52:102234
13. Alqaed S (2022) Effect of annual solar radiation on simple façade, double-skin facade and double-skin facade filled with phase change materials for saving energy. Sustain Energy Technol Assess 51:101928
14. Mukram TA, Daniel J (2022) A review of novel methods and current developments of phase change materials in the building walls for cooling applications. Sustain Energy Technol Assess 49:101709
15. Olia H et al (2019) Application of nanofluids in thermal performance enhancement of parabolic trough solar collector: state-of-the-art. Appl Sci 9(3):463
16. Sarafraz M et al (2019) Smart optimization of a thermosyphon heat pipe for an evacuated tube solar collector using response surface methodology (RSM). Physica A 534:122146
17. Young M et al (2020) Assessment of peak power demand reduction available via modulation of building ventilation systems. Energy and Buildings 214:109867
18. Zhang TT et al (2016) The application of air layers in building envelopes: a review. Appl Energy 165:707–734
19. Maghrabie HM et al (2022) A review of solar chimney for natural ventilation of residential and non-residential buildings. Sustain Energy Technol Assess 52:102082
20. Khanal R, Lei C (2011) Solar chimney—a passive strategy for natural ventilation. Energy Build 43(8):1811–1819
21. Cuce E et al (2019) Sustainable ventilation strategies in buildings: CFD research. Sustain Energy Technol Assess 36:100540
22. Khosravi R et al (2021) Entropy generation of graphene–platinum hybrid nanofluid flow through a wavy cylindrical microchannel solar receiver by using neural networks. J Therm Anal Calorim 145(4):1949–1967
23. Peng Y et al (2020) Investigation of energy performance in a U-shaped evacuated solar tube collector using oxide added nanoparticles through the emitter, absorber and transmittal environments via discrete ordinates radiation method. J Therm Anal Calorim 139(4):2623–2631
24. Omrani S et al (2017) Natural ventilation in multi-storey buildings: design process and review of evaluation tools. Build Environ 116:182–194
25. Shi L (2019) Impacts of wind on solar chimney performance in a building. Energy 185:55–67
26. Shi L et al (2019) Interaction effect of room opening and air inlet on solar chimney performance. Appl Therm Eng 159:113877–113888
27. Shi L (2018) Theoretical models for wall solar chimney under cooling and heating modes considering room configuration. Energy 165:925–938
28. Park D, Battaglia F (2015) Application of a wall-solar chimney for passive ventilation of dwellings. J Solar Energy Eng-Trans ASME 137(6)
29. Kong J, Niu J, Lei C (2020) A CFD based approach for determining the optimum inclination angle of a roof-top solar chimney for building ventilation. Sol Energy 198:555–569
30. Khanal R, Lei CW (2015) A numerical investigation of buoyancy induced turbulent air flow in an inclined passive wall solar chimney for natural ventilation. Energy Build 93:217–226

31. Shi L et al (2018) Determining the influencing factors on the performance of solar chimney in buildings. Renew Sustain Energy Rev 88:223–323
32. Li YG (2000) Buoyancy-driven natural ventilation in a thermally stratified one-zone building. Build Environ 35(3):207–214
33. Andersen KT (1995) Theoretical considerations on natural ventilation by thermal buoyancy. American society of heating, refrigerating and air-conditioning engineers
34. Cheng X et al (2020) Solar chimney in tunnel considering energy-saving and fire safety. Energy 210:118601
35. Tao Y et al (2021) Predicting airflow in naturally ventilated double-skin facades: theoretical analysis and modelling. Renew Energy 179:1940–1954
36. Baxevanou C, Fidaros D (2017) Numerical study of solar chimney operation in a two story building. Sustain Synergies Build Urban Scale 38:68–76
37. Asadi S et al (2016) The effect of solar chimney layout on ventilation rate in buildings. Energy Build 123:71–78
38. Zhang TT, Yang HX (2019) Flow and heat transfer characteristics of natural convection in vertical air channels of double-skin solar facades. Appl Energy 242:107–120
39. Yang D, Li P (2015) Natural ventilation of lower-level floors assisted by the mechanical ventilation of upper-level floors via a stack. Energy Build 92:296–305
40. Acred A, Hunt GR (2014) Stack ventilation in multi-storey atrium buildings: a dimensionless design approach. Build Environ 72:44–52
41. Axley J (1998) Introduction to the design of natural ventilation systems using loop equations. In: Document-air infiltration centre aic proc. Oscar Faber PLC
42. Axley J (2007) Multizone airflow modeling in buildings: history and theory. HVAC&R Res 13(6):907–928
43. Emmerich SJ, Dols WS, Axley JW (2001) Natural ventilation review and plan for design and analysis tools. US department of commerce, technology administration, national institute of…
44. Acred A, Hunt GR (2014) A simplified mathematical approach for modelling stack ventilation in multi-compartment buildings. Build Environ 71:121–130
45. Shi L, Zhang G (2016) An empirical model to predict the performance of typical solar chimneys considering both room and cavity configurations. Build Environ 103:250–261
46. McGrattan K et al (2013) Fire dynamics simulator user's guide. NIST special publication 1019(6)
47. Durrani F (2013) Using large eddy simulation to model buoyancy-driven natural ventilation. Loughborough University
48. Durrani F, Cook MJ, McGuirk JJ (2015) Evaluation of LES and RANS CFD modelling of multiple steady states in natural ventilation. Build Environ 92:167–181
49. McGrattan K et al (2013) Fire dynamics simulator technical reference guide volume 1: mathematical model 1018(1):175
50. Obyn S, van Moeseke G (2015) Variability and impact of internal surfaces convective heat transfer coefficients in the thermal evaluation of office buildings. Appl Therm Eng 87:258–272
51. McGrattan KB et al (2000) Fire dynamics simulator—technical reference guide. National institute of standards and technology, building and fire research…
52. Bouchair A (1989) Solar induced ventilation in the Algerian and similar climates. University of Leeds
53. Miao L, Chow CLJATE (2018) A study on window plume from a room fire to the cavity of a double-skin façade. 129:230–241
54. Tafakkori R, Fattahi A (2021) Introducing novel configurations for double-glazed windows with lower energy loss. Sustain Energy Technol Assess 43:100919
55. Cheng XD et al (2018) Study on optimizing design of solar chimney for natural ventilation and smoke exhaustion. Energy Build 170:145–156
56. Zhang H et al (2021) A wall solar chimney to ventilate multi-zone buildings. Sustain Energy Technol Assess 47:101381
57. Yang D, Li P (2015) Dimensionless design approach, applicability and energy performance of stack-based hybrid ventilation for multi-story buildings. Energy 93:128–140

58. Lee KH, Strand RK (2009) Enhancement of natural ventilation in buildings using a thermal chimney. Energy Build 41(6):615–621
59. Bassiouny R, Koura NSA (2008) An analytical and numerical study of solar chimney use for room natural ventilation. Energy Build 40(5):865–873
60. Burek SAM, Habeb A (2007) Air flow and thermal efficiency characteristics in solar chimneys and Trombe Walls. Energy Build 39(2):128–135
61. Shi L et al (2016) Developing an empirical model for roof solar chimney based on experimental data from various test rigs. Build Environ 110:115–128
62. Wang Q et al (2022) Solar chimney performance in buildings under three heating modes: an empirical analysis. Sustain Energy Technol Assess 52:102222
63. Yang LN, Xu PC, Li YG (2006) Nonlinear dynamic analysis of natural ventilation in a two-zone building: part a—theoretical analysis. Hvac&R Res 12(2):231–255
64. Wang Y, Wei C (2020) Nonlinear dynamic analysis of solution multiplicity of buoyancy ventilation in two vertically connected open cavities with unequal heights. complexity
65. Kumar R et al (2021) Experimental investigation of impact of the energy storage medium on the thermal performance of double pass solar air heater. Sustain Energy Technol Assess 48:101673

Chapter 7
A Mathematical Model for SC-Assisted Stack Ventilation in Multi-storey Buildings

Nomenclature

$A_{\text{in},i}$	Vent area of the ith floor, m^2
$A_{\text{w},i}$	Window area of the ith floor, m^2
A_{out}	Area of chimney top vent, m^2
A_{r}, i	Combined vent area ith floor, m^2
$A_{\text{s},i}$	Effective heating area of ith floor
B_i	Buoyancy flux of ith floor
C_{d}	Discharge coefficient
C_{p}	Specific heat capacity of ambient air, J/kg K
E_{c}	Heat input of each floor, w
f	Total number of floors of the building
g	Gravitational acceleration, m/s^2
$h_{\text{in},i}$	Air inlet height, m
$h_{\text{w},i}$	Window height, m
h_{sc}	Total height of building/solar chimney, m
q	Heat flux intensity
T_0	Environment temperature, K
T_{c}	Cavity air temperature, K
$V_{\text{in},i}$	A buoyancy-driven ventilation flow rate of the ith air inlet, m^3/s
$V_{\text{w},i}$	The ventilation flow rate through the ith window, m^3/s
$V_{\text{out},i}$	The ventilation flow rate through the chimney top vent, m^3/s
V_i	A modified buoyancy-driven ventilation flow rate of the ith air inlet, m^3/s

Greek Symbols

ρ_0	Density of ambient air, kg/m^3

L. Shi and H. Zhang, *Solar Chimney Applications in Buildings*, Green Energy and Technology, https://doi.org/10.1007/978-3-031-45218-5_7

ρ_c	Density of cavity air (kg/m^3)
τ	Transmissivity
α_i	Volume flow rate correction factor for each floor
κ_i	An averaged storey correction coefficient

Subscript

i	Floor number (*i*th)
0	Ambient
in	Inlet vent
out	Outlet vent

7.1 Introduction

Buoyancy-driven ventilation has been widely studied as an essential role in natural ventilation. By leveraging the potential for solar radiation, the passive solar design strategy could achieve multiple functions to reduce energy consumption [1, 2]. Following the airflow pattern, the passive ventilation system can achieve natural ventilation, space heating/cooling, and preheating air [3–6]. Air density gradients mainly produce pressure differences at different locations [7]. Buoyant ventilation systems featuring air channels, such as solar chimneys, Trombe walls, and double skin façades, share similar physical processes. Significantly, passive solar designs are preferable in tropical climates and medium/low-rise buildings [8, 9]. Besides, buoyancy-driven ventilation systems are adequate for low floors, while improving ventilation on high floors requires adjusting the vent area or using mixed modes of ventilation [10, 11].

The solar chimney is one of the most representative buoyancy-driven ventilation systems, mainly using passive solar energy to enhance ventilation. The induced airflow is accelerated within an openable air channel and the adjacent occupant spaces [12]. Extensive literature has explored the correlation between the ventilation capacity of solar chimneys and design parameters. The bulk works of Khanal and Lei [13–16] have focused on optimizing geometric parameters to improve SC's ventilation rates. Ren et al. [17] explored the flow characteristics and heat transfer mechanism of buoyancy within the air channel and found the positive effect of discrete heat sources on volume flow and heat transfer rates. Shi [18] developed four theoretical models for natural ventilation in solar chimneys for space cooling/heating, with quantitative results proven acceptable through experimental validations.

In the actual building environment, the influence of external wind is also considerable [19]. Shi [20] analyzed numerically and theoretically the influences of wind at different incident angles on the performance of SC and determined the critical wind speed. Jiménez-Xamán et al. [21] numerically simulated the thermal performance of the SC integrated into a single room in Mexico in summer and winter. Ventilation rates can be increased by 8–45% and 1.16–24.89% in these two seasons. Villar-Ramos et al. [22] experimentally verified that a single-channel solar chimney could achieve a maximum temperature difference of 16.7 °C and an air change rate of 27.11 h^{-1} by setting the air gap of the air channel to 0.175 m and the inclination angle of 15° under warm and humid climate conditions. However, most theoretical models and laboratory measurements reported in the literature were limited to the single-room/storey or an independent air channel.

Natural ventilation of multi-storey buildings generated by solar chimneys is still in the early stages of the investigation. There is no consensus on the theoretical model of multi-storey buildings with solar chimneys. A few studies on solar chimneys have highlighted evidence that naturally ventilated multi-storey buildings with solar chimneys can provide the required ventilation rate for each zone/floor. Punyasompun et al. [23] found that the air channel connected up and down was better than the separated air channel in reducing the indoor temperature. Another important finding from this study was that the numerical model was derived from the aspects of temperature and velocity by establishing a set of the thermal energy balance of the thermal network assisted by an electric analogy for a three-storey solar chimney building. A three-dimensional computational fluid dynamic (CFD) model was verified by experimental results and analytical models that tested the ventilation effect of a slanted roof solar chimney integrating into a two-storey building [24]. It was concluded that the results obtained from the 2D simulation understate the mass flow rate compared to the 3D prediction. While some common nature of the structure of the buoyancy-driven ventilation system with the air layer has been demonstrated in the literature, an understanding of the mechanism of the flow formation process and the ventilation effect of multi-storey solar chimney buildings remains elusive, and the evidence for related theories and practical methods was inconclusive.

The buoyant airflow from the air inlets of different floors converges and interacts in the air channel, complicating the airflow paths and thermal behavior. It highlights the significance of establishing a new theoretical model that can incorporate key influencing parameters, which is also easy to apply for solar chimney implementation in multi-storey buildings. In previous work, Gontikaki et al. [25] conducted dynamic simulation and sensitivity analysis of multi-storey office buildings and found that the solar chimney's dimensions and component material properties are influential parameters. Asadi et al. [26] used EnergyPlus to compare the ventilation rates when solar chimneys were placed in different parts of a seven-storey office building in Isfahan. Acred and Hunt [11] developed a simplified mathematical method for modeling buoyant flow behavior via flow governing equations formed by closed-loop pressure balance equations in an example three-storey building with a heated atrium and plenum. The developed equations were solved numerically with equal heat inputs on each storey. It is emphasized that only stack pressure and pressure loss at vents are considered in the pressure term. Air density stratification in each building zone is ignored to simplify the model.

However, most theoretical modeling and numerical approaches to buoyant flow characteristics were concentrated on the single-zone solar chimney or multi-storey atrium buildings, without considering the influence of floor quantities. This is due to the complicated situation in which different areas of the absorption wall drive the convergence of airflow from various rooms. Therefore, although the numerical simulations can predict the airflow rate through the different floors, the related optimization designs are very complicated and time-consuming as the associated impacts of the design factors on the performance are still unclear, not to mention a large number of the design factors for a solar chimney. It remains challenging to describe the flow transition and heat transfer with multiple inlets and a single outlet in multi-storey buildings, and it deserves additional exploration [27, 28].

Although running simulation software based on Computational fluid dynamics (CFD) can solve more complex fluid problems, large-scale multi-storey building cases require a lot of time and computing power. This study discusses ventilation performance numerically and develops empirical solutions to fill the gap. This work aims to establish a generalized theoretical model for estimating the ventilation rates of a multi-storey building equipped with a solar chimney, which can be solved by hand to reduce design costs. The wind environment is quite different at various locations. In this study, to offer a benchmark assessment of solar chimneys, the design of solar chimneys was carried out without considering the wind [29]. The three primary aims of this study are as follows:

- Analyze the performance of the solar-chimney-based stack ventilation in multi-storey buildings;
- Address the impacts of the leading influential factors on the natural ventilation mode; and
- Develop a theoretical model on the ventilation flow rate in multi-storey buildings with a solar chimney.

7.2 Methodology

7.2.1 *Theoretical Methodology*

According to previous studies, obtaining analytical solutions for nonlinear cases with three or more stories is difficult as the complexity and number of equations increase. Therefore, this section explores the solutions by establishing a theoretical model. The air in the air channel was assumed well-mixed and uniform with constant cavity temperature and air density. It is worth noting that the 'storey number' defined in this study represents the total number of floors owned by different buildings, while 'floor number' means the individual floor in a single building unless specified.

Due to buoyancy-driven ventilation's air flow and temperature behaving as nonlinear characteristics and solution multiplicity, predicting the ventilation performance is difficult, especially for large-scale naturally ventilated buildings [30, 31].

The airflow from each indoor building into the chimney cavity is heated and collected, then flows to the outdoors through the outlet. A series of flow equations composed of mass conservation, pressure balance equations, and the related equations between flow rate and flow rate were established to describe the flow process. Although the mathematical model has been simplified to allow analytical solutions to flow equations for buildings below two stories, the exact analytical solutions to the nonlinear equations become impossible with the number of stories (i.e., the air inlet increases). This study proposes a method of dividing each floor into independent calculations to quickly calculate the ventilation rate of each floor for multi-storey solar chimney buildings. The physical model treats air inlets independently and establishes flow equations separately, as shown in Fig. 7.1.

According to the closed flow loop analysis method [11], the process of air induced into the room from a window of each floor and then discharged through the outlet is conserved in mass. Fulfillment of the equation of conservation of continuity under steady-state conditions:

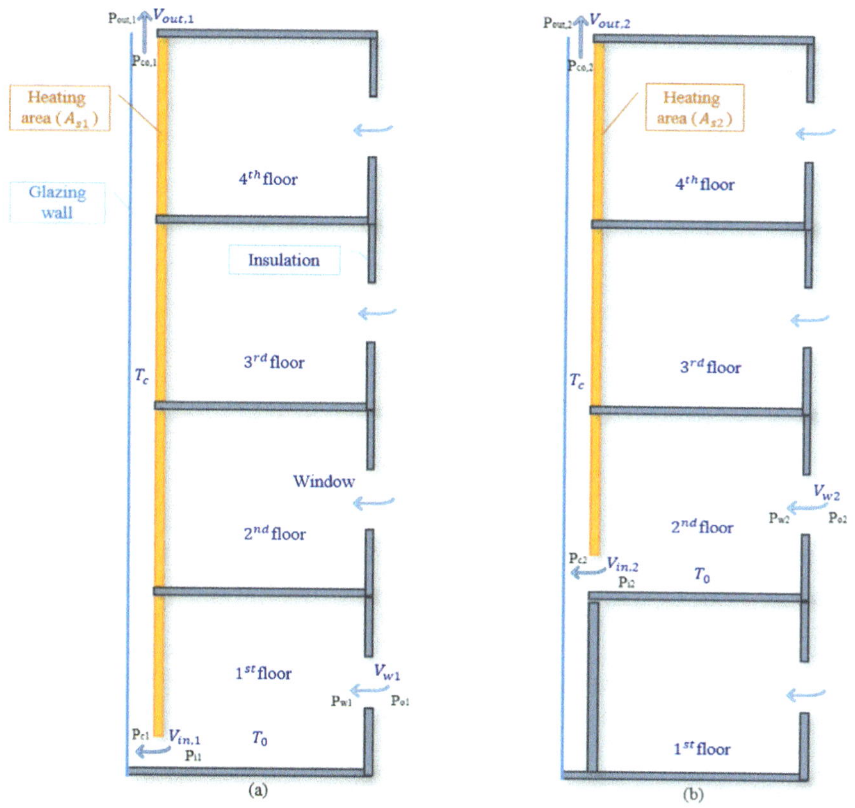

Fig. 7.1 An assumption of an example four-storey building under mathematical modeling: **a** the 1st floor; **b** the 2nd floor; **c** the 3rd floor; and **d** the 4th floor

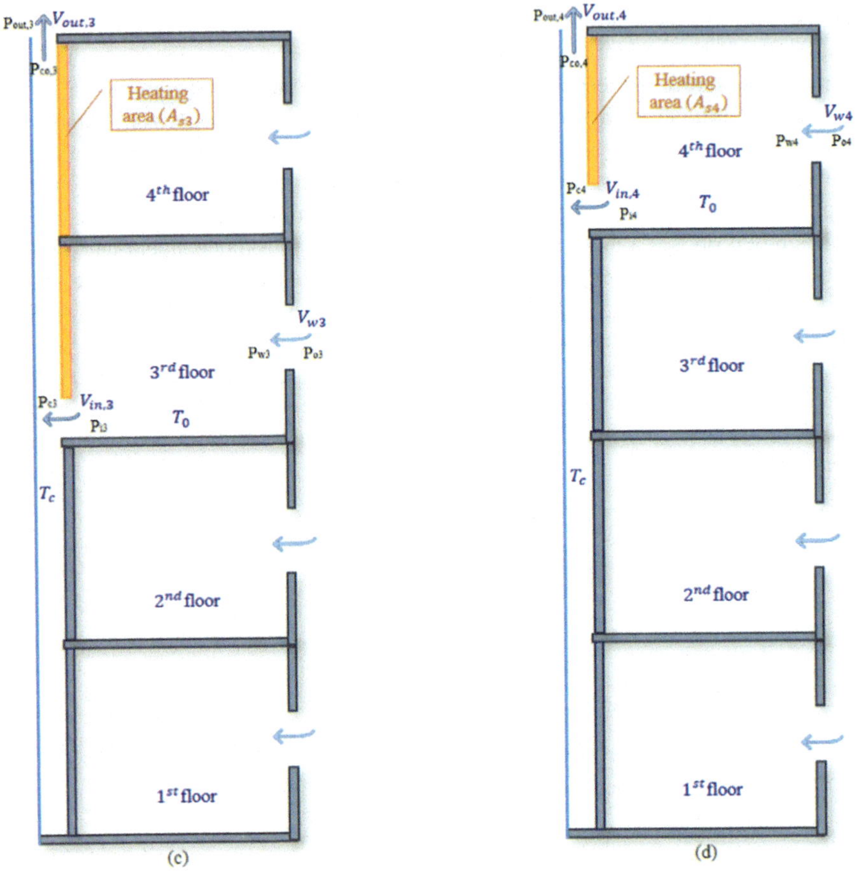

Fig. 7.1 (continued)

$$\rho_0 V_{\text{in},i} = \rho_0 V_{\text{w},i} = \rho_\text{c} V_{\text{out},i} \tag{7.1}$$

where subscript in represents the air inlet (or chimney inlet), subscript w stands for the window, and subscript i denotes the number of the vent. $V_{\text{out},i}$ denotes the discharge flow rate of each floor; subscript 'in' represents the air inlet; and ρ_c stands for the average air density within the solar chimney cavity.

Assuming that the air in the rooms and the chimney cavity are evenly mixed, the vertical momentum equation can be established, leading to the aerostatic pressure distribution at different heights inside the building [18]:

$$\Delta P = \rho g \Delta h \tag{7.2}$$

where ΔP stands for the pressure difference in the chimney cavity, room, and ambient air, Pa; Δh denotes vertical altitude difference, m; ρ is the air density, kg/m^3, and g denotes the gravitational acceleration, m/s^2.

Solar energy is used to heat the chimney cavity air and bring it to a constant temperature at a steady state. The conservation of energy leads to the following equation:

$$E_c = Q_s A_{s,i} = q\tau A_{s,i} = \rho_0 C_p V_{in,i} \Delta T \tag{7.3}$$

where Q_s denotes the net heat input; $\Delta T = T_c - T_0$; q denotes the heat flux intensity; E_c identifies the heating intensity power in the chimney cavity, W; τ denotes the glazing wall transmissivity, and $A_{s,i}$ denotes the effective heating area; C_p denotes the specific heat capacity of air, J/kg K.

As a supplementary equation, the relationship between the inlet pressure difference and flow velocity through a ventilation vent can be given by the modified Bernoulli equation [32]:

$$\Delta P_j = \frac{1}{2}\rho_j \mu_j^2 = \frac{\rho_j}{2}\left(\frac{V_j}{C_d A_j}\right)^2 \tag{7.4}$$

where ρ denotes air density, kg/m^3; ΔP denotes the pressure difference, N/m^2; μ denotes the air velocity through the aperture (i.e., window, air inlet/outlet), m/s; C_d represents the discharge coefficient; V stands for the volume flow rate via an aperture, m^3/s; A denotes the aperture area, m^2; and j represents the aperture number.

According to the assumption that indoor airflows according to a single path, as shown in Fig. 7.1, the flow path of each floor independently forms a closed flow loop. The relationship between $P_{o,i}$ and $P_{out,i}$ can be obtained following Eq. (7.2). The pair pressures differences on the two sides of the orifices: $P_{o,i}$ and $P_{w,i}$, $P_{c,i}$ and $P_{i,i}$, $P_{co,i}$ and $P_{out,i}$ can be given following Eq. (7.4). According to the balance of the overall pressure difference and conservation of mass, the above series of equations are superimposed, and the following equations can be developed,

$$\rho_0 g\left(h_{sc} - h_{w,i}\right) = \frac{\rho_0}{2}\left(\frac{V_{in,i}}{C_d A_{w,i}}\right)^2 + \rho_0 g\left(h_{in,i} - h_{w,i}\right) + \frac{\rho_0}{2}\left(\frac{V_{in,i}}{C_d A_{in,i}}\right)^2 + \rho_c g\left(h_{sc} - h_{in,i}\right) + \frac{\rho_c}{2}\left(\frac{V_{out}}{C_d A_{out}}\right)^2 \tag{7.5}$$

To intuitively show the relationship between $V_{out,i}$ and $V_{in,i}$, Eq. (7.5) is rearranged as follows,

$$\rho_c V_{out,i}^2 \frac{1}{C_d^2 A_{out}^2} = 2g(\rho_0 - \rho_c)\left(h_{sc} - h_{in,i}\right) - \rho_0 V_{in,i}^2 \times \left(\frac{1}{C_d^2 A_{w,i}^2} + \frac{1}{C_d^2 A_{in,i}^2}\right) \tag{7.6}$$

Introducing the ideal gas law as follows [18],

$$\rho_0 T_0 = \rho_c T_c \tag{7.7}$$

Substituting Eq. (7.1) into Eq. (7.6),

$$V_{\text{in},i} = C_\text{d} \times \sqrt{\frac{2g(\rho_0 - \rho_c)(h_\text{sc} - h_{\text{in},i}) \times A_{\text{w},i}^2 A_{\text{in},i}^2 A_\text{out}^2}{\rho_0 \times \left(A_{\text{in},i}^2 A_\text{out}^2 + A_{\text{w},i}^2 A_\text{out}^2 + A_{\text{in},i}^2 A_{\text{w},i}^2\right)}} \tag{7.8}$$

By substituting Eqs. (7.6) and (7.7) into Eq. (7.8) and setting,

$$A_{\text{r},i} = \sqrt{\frac{2A_{\text{w},i}^2 A_{\text{in},i}^2 A_\text{out}^2}{A_{\text{w},i}^2 A_{\text{in},i}^2 + A_\text{out}^2 A_{\text{in},1}^2 + A_\text{out}^2 A_{\text{w},i}^2}} \tag{7.9}$$

$$V_{\text{in},i} = C_\text{d}^{2/3} A_{\text{r},i}^{2/3} \times \left(\frac{E_i g}{T_c \rho_0 C_\text{p}}(h_\text{sc} - h_{\text{in},i})\right)^{1/3} \tag{7.10}$$

According to Eq. (7.7), the density variations in whole computational models can be neglected. For simplicity, the assumption of $\rho_0/\rho_c = 1$ is made. Let B_i represent the buoyancy input, the driving force of the buoyant air in the chimney.

$$B_i = \frac{E_i\, g}{T_0\, \rho_0\, C_p} \tag{7.11}$$

Then Eq. (7.10) can be rearranged to get the following equation,

$$V_{\text{in},i} = \left(C_\text{d} \times A_{\text{r},i}\right)^{2/3} \times \left[B_i \times (h_\text{sc} - h_{\text{in},i})\right]^{1/3} \tag{7.12}$$

The constants used in the mathematical method to describe the physical process are shown in Table 7.1.

The derivation process of the above flow equations can refer to a flowchart in Fig. 7.2. This equation preliminarily determines the mathematical model of the volume flow rate of each floor for multi-storey solar chimney buildings. To obtain a

Table 7.1 Modeling inputs of the theoretical models [18]

Constants	Nomenclature	Value	Unit
C_d	Discharge coefficient	0.6	–
C_p	Air specific heat capacity	1005	J/kg K
ρ_0	Ambient air density	1.205	kg/m^3
g	Gravity acceleration	9.81	m/s^2
T_0	Ambient temperature	293.15	K
τ	Glass transmissivity	0.76	–

computational model closer to the actual situation, the revised mathematical model will be introduced in Sect. 7.3.3.

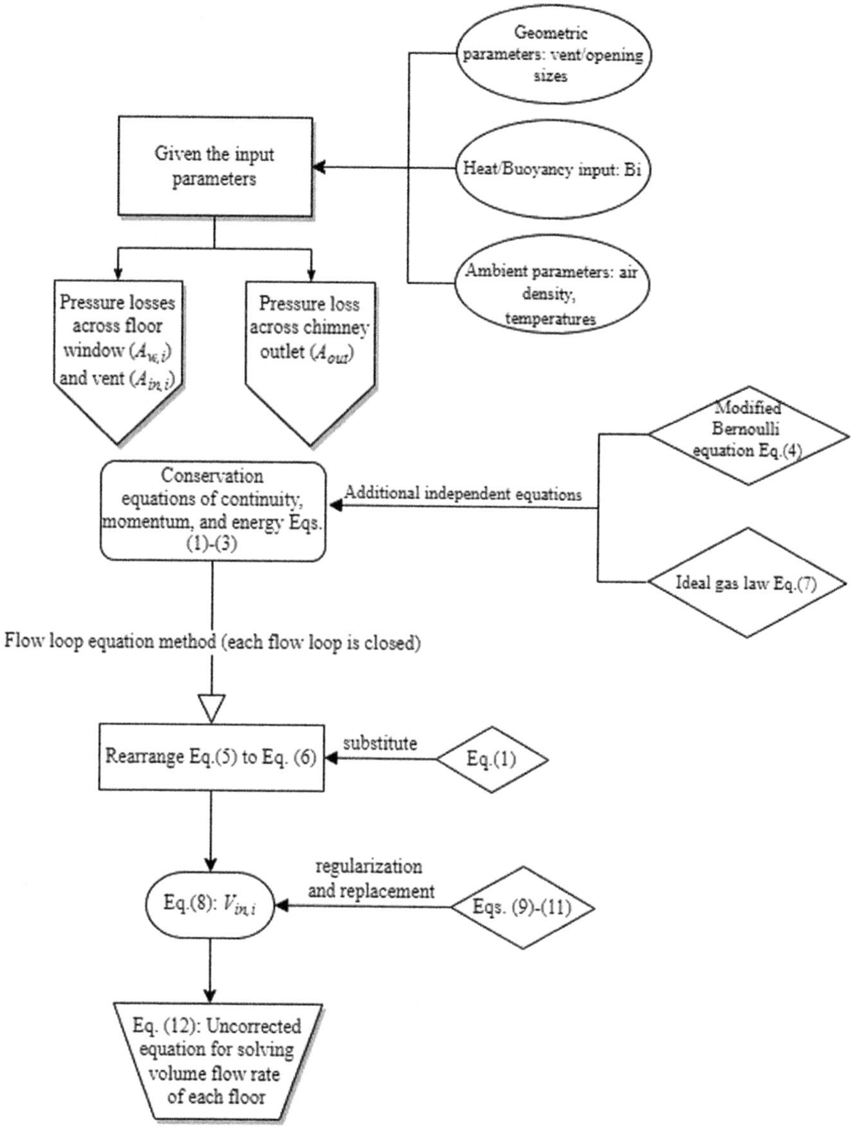

Fig. 7.2 Flowchart for the theoretical modeling

7.2.2 *Numerical Methodology*

7.2.2.1 Numerical Modeling

This section illustrates an example of a four-storey building (Fig. 7.3) heated by solar radiation. A vertical air channel (d) of the solar chimney is attached to the building envelope, consisting of a heat-absorbing wall and a glazing wall. All solid building surfaces were considered smooth and inert by default. A continuous and uniform heat flux heats the absorption wall to simulate radiation absorption. Individual zones are interconnected by flow paths and form an airflow network. The room on each floor communicates with the chimney cavity through a single air inlet. Each floor has identical inner dimensions of 6.0 m (L_r) × 6.0 m (W_r) × 3.0 m (H_r), a single air inlet (h), and a square window positioned in the center of each floor's left wall. The solar chimney's upper section has an air vent. The vertical distance between the ground and each air inlet center ($h_{in,i}$) represents the air inlet position. The position of each air inlet can be moved from the floor to the ceiling. h_{sc} represents the total height of a solar chimney and is equivalent to building height. w_c stands for the chimney width that is equivalent to the building width. In the reference case, dimensions include a solar chimney with a 0.3 m air channel/cavity gap (d), a 0.3 m air inlet height (h), a 6.0 m chimney width (w_c), and a chimney height of 12.6 m. The size of room windows is 1.0 m × 1.0 m (A_{open}).

Table 7.2 shows the variables considered in the numerical study. The selected parameters that affect the ventilation performance of solar chimneys build on evidence from prior studies [33], including the geometric configurations and the external environmental factor. This study evaluated a total of 35 scenarios by adjusting a single design parameter simultaneously, with scenario No. 0 as a benchmark case.

7.2.2.2 Numerical Approach

This study employed FDS based on the large eddy simulation (LES) model to solve the energy equations and predict the volume airflow rates driven by given heat fluxes for multi-storey solar chimney buildings. The software FDS was developed by the U.S. National Institute of Standards and Technology (NIST), with credits and reputation. The software numerically solves a form of the Navier–Stokes equations appropriate for low-speed, thermally driven flow, emphasizing smoke and heat transport. Previous studies have confirmed the effectiveness and accuracy of FDS for natural ventilation predictions [34–36]. The continuity, momentum, and energy conservation equations refer to our previous studies [37].

The continuity equation is given by,

$$\frac{\partial \rho}{\partial t} + \nabla \cdot \rho \vec{u} = 0 \tag{7.13}$$

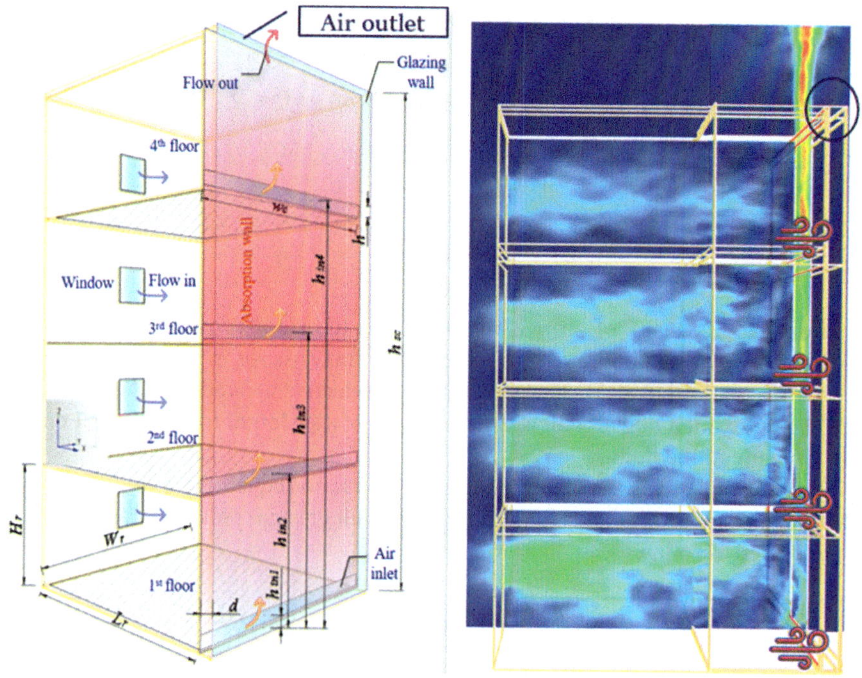

Fig. 7.3 A schematic diagram of an example four-storey building with a solar chimney

Table 7.2 An overview of the scenarios that were studied concerning the different affecting factors

Influencing parameters		No	Values	Interval	Unit
Benchmark case		0	Factors[a]		
Chimney	Inlet size (h_{in})	1–4	0.1–0.5	0.1	m
	Inlet location (h_l)	5–13	0.15–2.85	0.3	m
	Cavity gap (d)	14–20	0.1, 0.2, 0.5, 0.8, 1.0, 1.2, 1.5	–	m
Room	Window area (A_{open})	21–24	1.2 × 1.2, 1.5 × 1.5, 1.8 × 1.8, 2.0 × 2.0	–	m^2
External condition	Radiation intensity (q)	25–29	200–1000	200	W/m^2
Floors	Number of storeys	30–35	1–7	1	–

Notes [a] The benchmark scenario serves as the foundation for the other scenarios, namely a four-storey building with each floor area of 6 m × 6 m and 3 m height; inlet size: 0.3 m; inlet location: 0 m; cavity gap: 0.3 m; chimney height: 12.6 m; window area: 1.0 m square; solar intensity: 500 W/m^2. One influential parameter changes each time, while the other values stay constant with the benchmark case

The momentum equation is expressed as,

$$P\left(\frac{\partial \vec{u}}{\partial t}+(\vec{u}\cdot\nabla)\vec{u}\right)+\nabla p=\rho g+f+\nabla\cdot\vec{\tau} \tag{7.14}$$

The energy equation is written as,

$$\frac{\partial}{\partial t}(\rho h)+\nabla\cdot\rho h\vec{u}=\frac{Dp}{Dt}-\nabla\cdot\mathbf{q}+\varepsilon \tag{7.15}$$

where ρ is the density of air, t is the time, $\vec{u}$ is the velocity vector (in x-, y-, and z-directions), p is the pressure, f is the external force vector, $\vec{\tau}$ is the viscous stress tensor, h is the enthalpy, $\nabla\cdot\mathbf{q}$ represents the conductive and radiative heat flux, and ε is the dissipation rate at which kinetic energy is transferred to the thermal energy via viscosity of the fluid.

Thermal radiation's effect on the energy conservation equation was modeled using Deardorff's model. An algorithm that uses a default grey gas radiation transfer model is expressed as follows [38]:

$$s\cdot\nabla I_\lambda(x,s)=\kappa(x,\lambda)[I_b(x)-I(x,s)] \tag{7.16}$$

where $I_b(x)$ denotes a Planck's function source term, $I_\lambda(x,s)$ denotes the radiation intensity at wavelength λ, $\kappa(x, \lambda)$ denotes the absorption coefficient at an x of the wavelength λ, and s denotes the unit normal direction vector.

FDS can simulate natural ventilation scenarios using computational grids as an efficient flow-solving technique based on LES for steady and unsteady airflows [39], which filters out small eddies while directly solving conservation equations for larger energy-containing eddies [40]. The LES approach accomplishes the scale separation by a low-pass filter of width Δ—the cube root of the cell volume ($\Delta = V_c{}^{1/3}$, where $V_c = \mathrm{d}x\mathrm{d}y\mathrm{d}z$), and captures the primary flow features that are larger than the width. For an arbitrary continuous field, Φ, a filtered field is defined as:

$$\Phi(x,y,z,t)=\frac{1}{V_c}\int_{x-\frac{\delta x}{2}}^{x+\frac{\delta x}{2}}\int_{y-\frac{\delta y}{2}}^{y+\frac{\delta y}{2}}\int_{z-\frac{\delta z}{2}}^{z+\frac{\delta z}{2}}\phi\left(x',y',z',t\right)\mathrm{d}x'\mathrm{d}y'\mathrm{d}z' \tag{7.17}$$

The building walls are well insulated, and heat conduction between the wall and the air is not considered. According to the solar chimney's working mechanism under ventilation mode, the numerical model in this study operates with a constant flow of fresh air, supplying and ensuring unidirectional flow. It implies that the indoor air temperature is infinitely close to the ambient, and only the convective heat flux in the air channel drives the airflow and produces thermal stratification.

FDS has a default model for describing convective heat transfer between the heated wall and ambient air [40]. Given that heat flux powers the buoyant effect, the presence of a heated absorption wall makes the cavity temperature consistently higher than that of the rooms. Except for the heated absorption wall, all solid surfaces are considered to have smooth and inert boundary conditions by default to speed up numerical convergence. This study specifies the convective heat transfer coefficient as 15.0 W/(m^2 K) [41], and the glazing wall has a transmittance of 0.76 [20]. The values for pressure and ambient temperature are 1 atm and 20 °C.

Three grid sizes are used in the numerical simulation with the four-storey building benchmark case as an example to examine grid independence. Figure 7.4 shows the total volume flow rate at the chimney outlet under grid sizes of 0.10 m, 0.05 m, and 0.025 m, respectively. The flow rate decreases slightly when the grid size is reduced from 0.1 m to 0.025 m. When the mesh size is less than 0.05 m, the output flow difference is almost imperceptible when the simulation reaches a stable stage. The grid size of 0.05 m was found to be sufficiently accurate compared to further refined grid sizes. Also, given that the multi-storey building has a much larger domain than a single room, two mesh cell sizes of 0.1 m and 0.05 m are adopted to save calculation time. Where closer to the solar chimney—1 m distance within the air inlet, the mesh size is 0.05 m in three dimensions, while domains further away from the chimney air inlet gradually transition to 0.1 m.

A set of transient volume flow rates were yielded from the numerical simulation. The stack's flow characteristics were averaged across the cross section of the vents. A time period of 100 s was determined to be sufficient to guarantee a stabilized outputted airflow. The amount of airflow through the air inlets and the outlet can be used to estimate the solar chimney's ventilation capacity. As a result, the volume

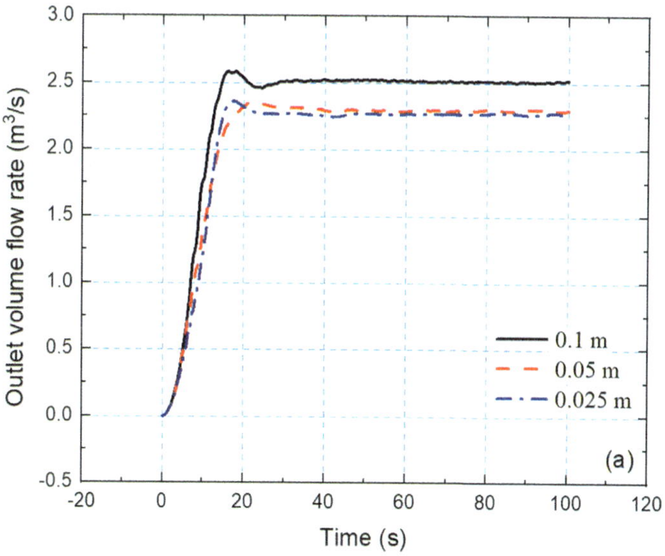

Fig. 7.4 Comparison of volume flow at outlet using different mesh sizes under 500 W/m^2 solar radiation. The X-axis shows the time, and Y-axis represents the outlet volumetric flow rate

flow rates for the last 30 s of the duration were averaged. The time step size is adjusted dynamically in solving discrete equations using the FDS solver to save computational time. The initial default time step size is determined automatically by mesh cell size and the flow velocity. Throughout each iteration, each mesh's time steps are synchronized. More detailed information regarding the setting can be found in Ref. [42].

7.2.2.3 Validation of a Numerical Tool

The validation of the numerical simulation was confirmed by Shi et al. [43] according to a reduced-scale solar chimney experimental test platform with a room of 1.5 m × 1.5 m × 0.9 m built by Cheng et al. [44]. The experimental setup used an artificial light source to simulate uniform solar radiation heating in the range of 400–1200 w/m^2. It was noted that when numerical modeling is used in conjunction with natural ventilation, the findings are reasonably consistent with tests. However, the numerical predictions differ from the experimental results for the two extreme inlet positions (the bottom and the top of the absorption wall) due to the complicated turbulent mixture and significant pressure loss during airflow transfer from the window to the air inlet.

Moreover, another experimental test of the solar chimney was undertaken by Bouchair [45]. The heating device uniformly heated the absorption and glass walls at four temperatures (30, 40, 50, and 60 °C). Numerical simulations compared corresponding experiments under different scenarios, including two air inlet heights, four different surface temperatures, and five air gaps. The obtained numerical model (i.e., FDS) fits reasonably well with experimental data. Our earlier work examined the numerical and experimental findings of the mass flow generated by a solar chimney [35]. Despite the increasing discrepancy in numerical and experimental data under a wider cavity gap, the numerical results demonstrate consistency with the experimental results. However, a reasonable agreement between numerical models and experimental findings was observed, particularly in scenarios with a cavity gap of 0.1–0.3 m with a range of absolute error from 0.11 to 13.6%, while an average absolute error of 32.4% occurs at a cavity gap of 1.0 m. Also, an average absolute error of 21.1% between simulations and experiments was verified in [46] under natural ventilation. The accuracy of the numerical method was shown to be credible by a comparison of the numerical prediction with current empirical equations and full-scale experimental data [47].

Furthermore, the validation of FDS was also carried out in our previous study [20]. It is found that the numerically predicted volume flow rate through the chimney cavity fits reasonably well with experimental data under natural ventilation. Since the validity of the numerical model was confirmed in Refs. [20, 43], FDS was then used to perform the numerical modeling for this study. Due to the length limitation, the related validations were not repeated here, and more details can refer to our previous studies [20, 43].

7.3 Results and Discussion

7.3.1 Chimney and Building Configurations

7.3.1.1 Vertical Airflow Patterns

This study focuses on the flow pattern and volume flow rates throughout rooms into the air channel. The multiple flow regimes caused by the stack effect are demonstrated from an intuitive multi-dimensional perspective for a four-storey building. The solar radiation bursts into the vertical air channel through the glazing wall, hitting the absorption wall and warming up the air inside the air channel. Due to the pressure gradient, the stale air within the rooms enters the air channel via the inlets and flows out through the solar chimney's outlet. This study establishes different scenarios ranging from one-storey buildings to seven-storey buildings. Given the limited space, unless otherwise stated, the visualization simulation results in this article are taken as an example of a four-storey building.

Figure 7.5a in a side view shows the velocity distributions and flow pattern of an example four-storey building under a solar radiation intensity of 500 W/m^2. The vertical central plane of the building and the air channel was obtained from the FDS model. The buoyant flow is generated by passive solar, so the air supply rate varies each storey owing to the stack effect's magnitude. The buoyancy intensity lessens with increasing building height due to the solar chimney's decreased impact on stack pressure and buoyancy-induced airflows. Since the buoyancy effect diminishes as the level rises, the air velocity on the upper floors and the range of the high-speed zone is considerably smaller than those on the lower floors. A high-velocity region inside the air channel and a distinct difference in ventilation flow on every floor can be observed. The airflow forms a network through the flow path formed by intentionally designed vents.

Figure 7.5b in a front view shows the vertical movement of the airflow in the air channel as the floor increases. Unlike the rooms' air velocity distribution, the cavity air velocity progressively rises as the height increases. Because the air temperature in the air channel is higher than that of the surrounding areas, the airflow velocity in the air channel is considerably greater than that of the room. The air near the absorption wall is heated first to create a density difference with the surrounding air, achieving the maximum flow rate sooner. Furthermore, cavity airflow is the superposition of the airflow flowing into the air channel from each floor. Thus, to measure the amount of airflow contributed by each floor, the location for detecting the volume flow rate is at the interface of air inlets and the outlet.

7.3.1.2 Gap and Inlet Dimensions

The influences of the SC's dimensional parameters (inlet size and position, cavity gap) and building parameters on the volume flow rates are discussed in this section,

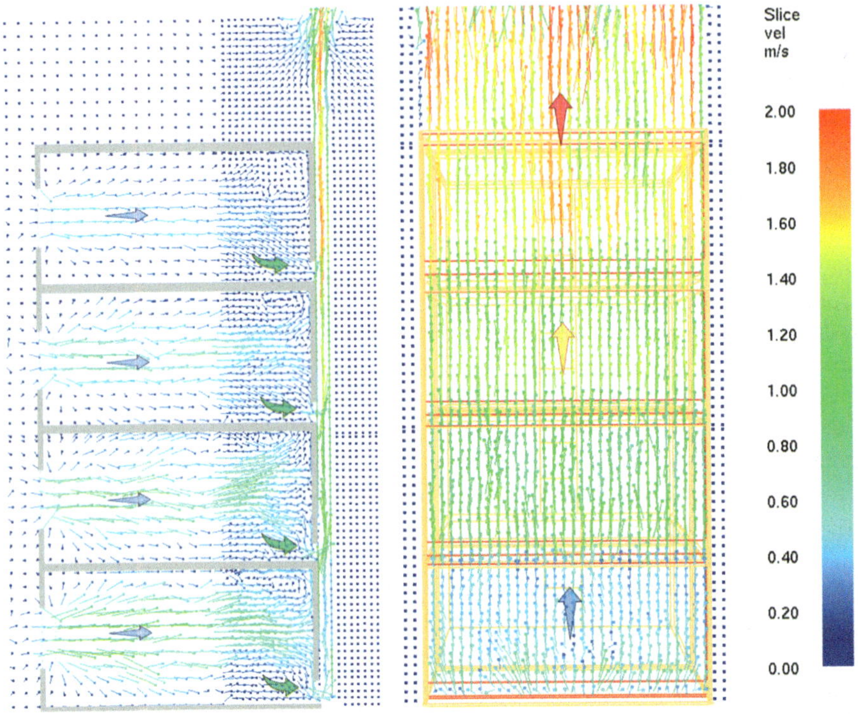

Fig. 7.5 Airflow velocity distribution for the four-storey case: **a** side elevation; and **b** central elevation plane of the air channel. On the right, a color bar indicates a velocity range of 0–2 m/s

taking the four-storey building as an example. Figure 7.6a shows that air inlet size (h) influences ventilation performance. Volume flow rates respond differently to changes in the vent sizes that are equivalent and simultaneous. The first- and fourth-floor volume flow rates are more sensitive to the air inlet size, while changes are imperceptible in the middle two floors. The volume flow rate gradually increases with the air inlet size for the 4th air inlet. However, the improving effect becomes less evident as the air inlet size exceeds 0.2 m. Differently, the changing tendencies of the 1st-floor flow rate are opposite to that of the fourth floor. When the air inlet size exceeds 0.2 m, the volume flow rate drops significantly.

Also, the outlet flow initially increases with the air inlet height and falls with the inlet height exceeding 0.2 m. A relatively low air inlet and a relatively high air outlet are conducive to buoyancy-driven natural ventilation. In other words, the greater the height difference between the air inlet and the air outlet, the better the buoyancy effect. The first floor already enjoys a superior location advantage in terms of obtaining maximum buoyancy. When the air inlet of the chimney is at the same level as the room window, the flow rate reaches an extreme value because the incoming airflow is less obstructed. Then, the buoyancy effect is suppressed due to the continuous reduction of the height difference between the air inlet and the air outlet. Similar to

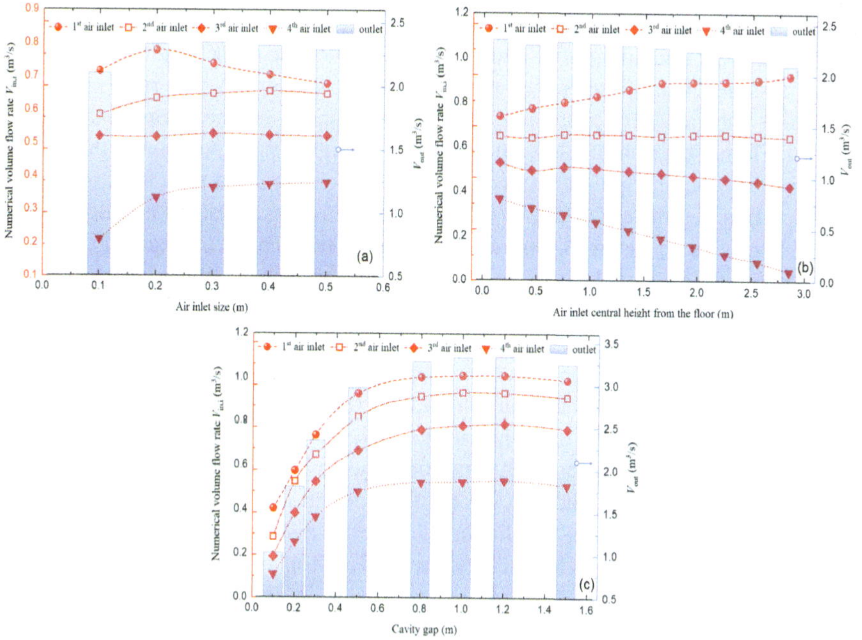

Fig. 7.6 Influences of chimney arrangement on volume flow rate: **a** inlet size, **b** inlet location, and **c** cavity gap. The flow rate at the air outlet is the summation of the airflow rates through all the air inlets

the bottom floor, the flow trend of the top floor gradually flattens out after a brief rise. The floor in the middle is less sensitive to the change in the air inlet position due to the slight change in stack height. The shift in driving stack pressure from buoyant cavity air is negligible.

Additionally, this study establishes numerical cases with the same configuration from a single-storey to a seven-storey building to explore the influential parameters corresponding to the peak volume flow rate when the floor quantities increase. It is found that as the air inlet height increases from 0.1 to 0.5 m, trends in data summarized all the cases from the single-storey to the seven-storey buildings indicate the air inlet sizes corresponding to the peak volume flow rate: 0.2 m air inlet height for single-storey and seven-storey, 0.4 m for two-storey, 0.3 m for three- to six-storey buildings. Overall, these findings are in accordance with findings reported in the single-zone case [44].

The variations in the volume flow rates with inlet heights are shown in Fig. 7.6b. The inlet height (h_{in}) moves upward along the absorption wall. There are significant discrepancies in the flow rate on each floor. As the relative positions of the air inlet and outlet continue to decrease, the weakening of the buoyancy effect has a different degree of influence on the flow rate of each floor. The 1st air inlet and 4th air inlet present opposite trends. The air inlet's upward movement for the fourth air inlet leads to the decreased stack effect. The 4th air inlet is the vent closest to the outlet, and its stack height is the shortest, so it is more sensitive to the difference in height between the air inlet and outlet. Due to the benefit of the stack height and heat wall area constantly being higher than the other floors, the airflow rate rises rather than decreases substantially for the inlet on the 1st floor. As for the floors in the middle of the building, the volume flow rate is hardly affected by the position of the air inlet because the stack height difference is not as evident as the top and bottom floors' changes. Besides, the efficient area with heat inputs is lowered as the air inlet upward, which harms the pressure difference of thermal buoyancy. Consequently, the total outlet volume flow rate drops from 2.35 to 2.10 m^3/s.

Figure 7.6c illustrates the varying trends of the flow rates caused by variations in air channel/cavity gap (d). The volume flow rate at all vents exhibits the same changing pattern. The volume flow initially experiences a steep increase and tends to be stable, with the cavity gap enlarging beyond 0.8 m. Subsequently, the volumetric flow exhibits a downward trend when the gap is larger than 1.2 m. Besides, according to the total numerical scenarios of this study, the cavity gap where the volume flow rate reaches the peak also increases as the floor quantities increase, such as 0.3 m for the single storey [8, 48–52] buildings, 0.5 m for two-storey, 1.0 m for three-storey, 1.2 m for four-storey and five-storey, and 1.5 m for six-storey and seven-storey buildings.

7.3.1.3 The Air Entrance Window

Figure 7.7 demonstrates that the improvement of volume flow rates is nearly independent of the window area. The increase in the window area is initially effective at increasing the volume flow rate of the 1st air inlet and outlet. However, the improving effect becomes less evident as the window area increases beyond 1.5 × 1.5 m. Differently, the volume flow rate of other floors almost remains at a constant value without significant fluctuation. Compared to the single-zone case reported in the literature [35], upper-floor volume flow rates appeared unaffected by enlarging window areas.

This study compares the influence of different storey numbers on the solar chimney's performance under benchmark scenario settings. The buoyancy-induced ventilation of each floor in a multi-storey building gradually decreases from the lower floor to the upper floor. Taking the example of a seven-storey building with solar chimneys, the flow rate on the 1st floor is 3.4 times that of the 7th floor.

Besides, if the chimney height is always kept flush with the building height, the expansion in floor amounts will undoubtedly increase the stack height. The increase in chimney height is beneficial for enhancing the thermal buoyancy effect. However, the growth rate of volumetric flow gradually decreases as the storey number increases. Figure 7.8 shows that as the total building height increases, the volumetric flow at the air inlets of each layer also increases accordingly. Similar to the evidence of Ref. [10], adding more levels may hinder the establishment of buoyant flows on lower-level floors. More floors mean that the channel's overall height is higher, which causes a longer retention time of the air. A large amount of airflow requires a sufficient pressure difference to discharge. Besides, as the flow of lower floors continues to increase,

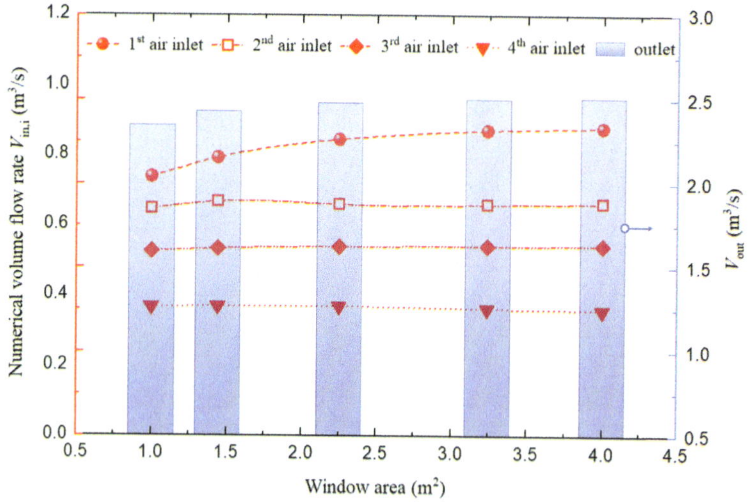

Fig. 7.7 Effects of window area on volume flow rates

the cavity temperature decreases, which is not conducive to the accumulation of buoyancy. Despite the total airflow rate at the chimney outlet increasing from 1.24 to 3.55 m^3/s (the values are read from the chimney outlet) as the building volume is increased from a two-storey building to a seven-storey building with the identical configuration and solar radiation intensity, the volume flow rate of the top floor of each building decreased exponentially: $V_{in,top} = 0.64e^{-0.18f}$, where f denotes storey number, as shown in Fig. 7.8.

7.3.2 *Solar Radiation*

Solar-assisted passive ventilation is heavily reliant on solar radiation intensity. Figure 7.9 presents how the volume flow rates vary in response to changes in solar radiation intensity. The volume flow rates exhibit a power function growth with the solar radiation intensity from 200 to 1000 W/m^2. The solar energy received by the lower floor is greater than that of higher floors due to the different heating areas, resulting in a higher volume flow rate on the lower floors.

The regression can explain the relationship between heat flux intensity and volume flow rates: $V \propto q^{1/3}$, where q denotes the solar radiation intensity, W/m^2. A similar power function relation $V \propto q^{1/2}$ was reported by Shi et al. [53]. From the value of the power in the regressions, it is found that as the volume and quantity of the ventilated space increase, the influential degree of solar radiation on the volume flow rate decreases slightly.

The exhaust air volume rate can reflect the overall performance of the solar chimney at the outlet (V_{out}). Figure 7.10 illustrates all numerical simulation scenarios

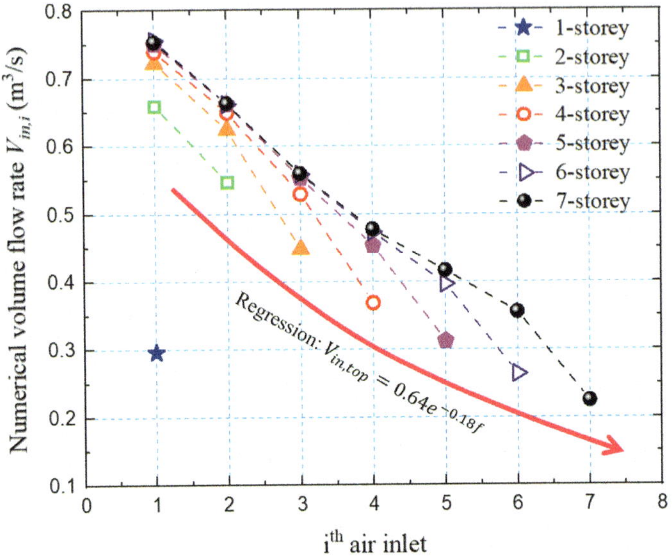

Fig. 7.8 Storey number impacts volume flow rates at the individual air inlet of each floor

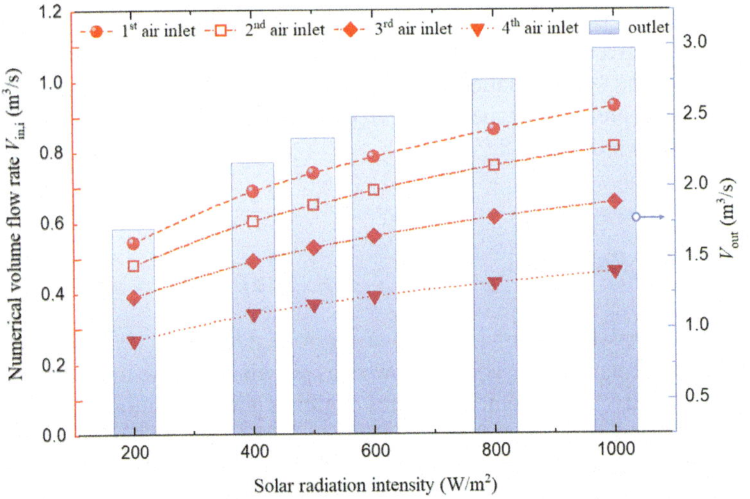

Fig. 7.9 Influence of heat input intensity on air inlet flow

with different floors (i.e., from single- to seven-storey buildings) established in this study. By comparing with the results of the corresponding reference case, the geometric parameter setting corresponding to the maximum volume flow can be obtained. The approach and the improvement degree of different buildings to obtain the maximum volume flow are as follows: 43% improvement in the volume flow rate

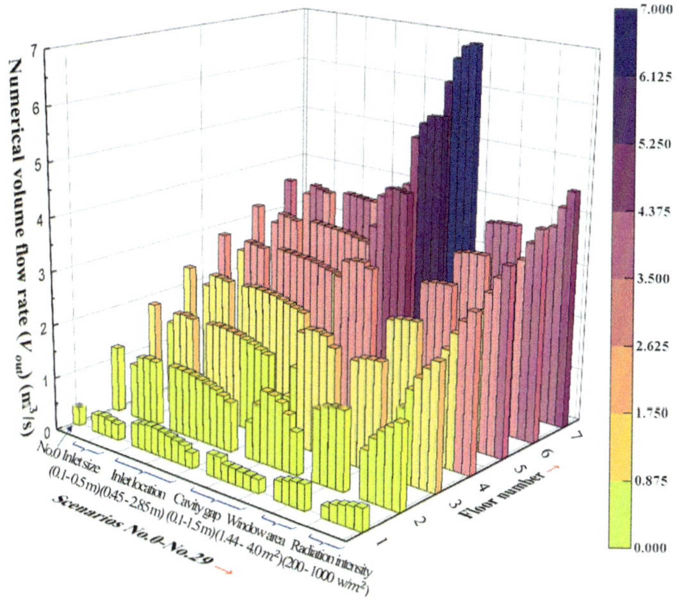

Fig. 7.10 Total volume flow rate (V_{out}) at the chimney outlet of all the scenarios from the single-storey to seven-storey buildings. The unit of the color bar is m^3/s

can be achieved by adjusting the position of the air inlet to be flush with the window for single floor; 26.4 and 27.9% increases occurred when solar radiation reached 1000 W/m^2 for two-storey and three-storey buildings, respectively; The range from 42.2 to 93.9% enhancement is achieved by increasing the cavity gap to a specific value for four-storey to seven-storey buildings.

For air channels to increase ventilation capacity, compared with the benchmark case (No. 0 in Table 7.2) for each building, designing the cavity gap corresponding to a peak value of the volume flow rate is the most effective strategy under a constant heat flux. It is worth noting that the improvement of the cavity gap on volume flow becomes more and more evident with the increase of floors. The second is the influence of the input heat fluxes. When the geometric parameters of the building and the solar chimney remain unchanged, the higher the heat flux, the more volume flow will be generated. The third is that enlarging the window area positively promotes the flow rates, but the continued increasing window area does not improve significantly. Due to the different responses of the volumetric flow rate of each floor to the inlet size and location, the aggregated overall volume flow rate discharged from the solar chimney has not been significantly improved.

7.3.3 Storey Correction Coefficient

It should be noted that the simplified theoretical model based on a traditional model [18, 46, 54] that the airflow in the air channel is regarded as well-mixed and uniform [18, 46] under given heat flux, suffering from a bias since the density difference caused by the vertical linear and horizontal parabolic distribution of temperature is ignored. The evidence errors obtained in the analytical method are biased upward, and the deviation increases with the increase of floors. Therefore, a storey correction coefficient (SCC) that changes with the floor quantities is proposed in this study. Denoting the storey correction coefficient between the numerical modeling and the analytical method as κ_i,

$$V_i = \kappa_i V_{\text{in},i} \tag{7.18}$$

where V_i is the volumetric flow rate at each air inlet obtained by numerical simulation; and κ_i represents the average value of each storey correction coefficient for a type of building (refer to each column in Table 7.3); $V_{\text{in},i}$ is the value calculated using Eq. (7.12) according to the hypothetical scenario in Fig. 7.1.

What stands out in Table 7.3 is each storey correction coefficient α_i to define the influence of the difference in the storey number through linear regression.

Figure 7.11 presents the comparison of numerical evidence and empirical results. By introducing averaged storey correction coefficient (κ_i) from regression, the acceptable estimated results can be obtained. In the figure, the X-axis is the volume flow obtained by numerical simulation, and the Y-axis is the result calculated by Eq. (7.18). It can be seen from the 45° line that most of the data are consistent. The

Table 7.3 Storey correction coefficient of each floor was obtained by comparison with the numerical model

Each storey correction coefficient α_i, $\left(\frac{V_i}{V_{\text{in},i}}\right)$ for *i*th air inlet	Two-storey	Three-storey	Four-storey	Five-storey	Six-storey	Seven-storey
α_7						0.44
α_6					0.47	0.39
α_5				0.53	0.42	0.36
α_4			0.59	0.47	0.39	0.35
α_3		0.67	0.53	0.44	0.38	0.34
α_2	0.79	0.60	0.50	0.43	0.38	0.34
α_1	0.67	0.56	0.48	0.42	0.37	0.34
Average: κ_i	0.71	0.60	0.51	0.44	0.39	0.36

Note *i* denotes the number of floors; and κ_i represents the average value for each building (column)

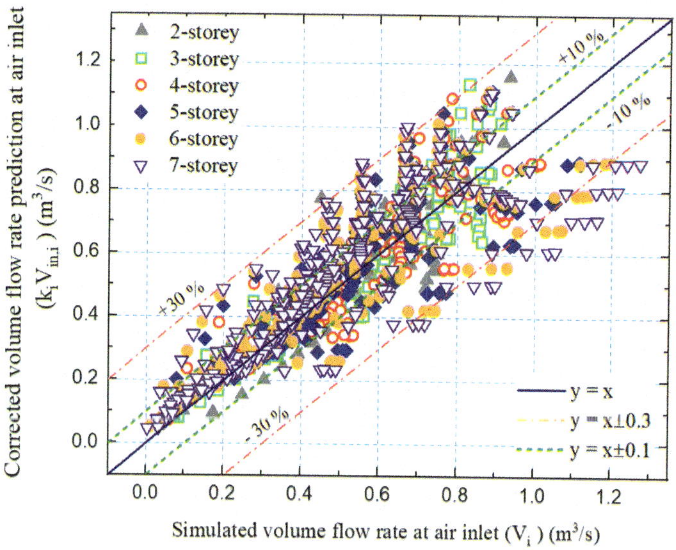

Fig. 7.11 Comparison of numerical results and empirical results

simulation results in some extreme scenarios will deviate more from the calculated value (within ±30%). In order to simplify the analysis and calculation, the reduction coefficient is integrated to obtain the average, which will inevitably lead to an integration error. Replacing the measured value with the group median value of each column (refer to Table 7.1), the root cause of the error resulting from the addition of the averaged reduction factor is individual variation. The greater the individual differences, the greater the error. For example, scenarios with an error greater than 30% are $V_{\text{in1}} \sim V_{\text{in4}}$ with cavity gap larger than 1.0 m for 6-storey and 7-storey. In addition, with the increase in floor quantities, the first floor and the top floor have a larger deviation from the middle floor due to the median value of the reduction coefficient. In other words, the empirical equation only applies to low-rise buildings.

From the error beyond ±10% and within ±30%, cases where the gap is less than 0.3 m and the room with a large opening show that the empirical equation will overestimate the volume flow rate of the air inlet on each storey. From the numerical simulation, it can be seen that the window size has a relatively small impact on the flow, and the impact on the middle and high floors is almost imperceptible. For simplicity, the area of each vent in the empirical equation uses the actual geometric area rather than the effective ventilation area, which increases the effect of $A_{w,i}$ and further led to the deviation of the empirical results from the numerical simulation. When the gap is greater than 0.5 m, the unheated area in the cavity increases, and flow reversals occur in the area close to the air outlet, resulting in more turbulent airflow in the cavity, which makes the empirical equation underestimates the flow rates on the upper floors. If the chimney is raised above the height of the roof, the flow at the air inlets on the floors below the top floor will be overestimated, and top floor flow will be underestimated. This is because increasing the chimney height weakens the buoyancy effect of the lower floor, while the effect on the top floor is similar to that of a single-zone SC building, which increases the pressure difference and thus the flow rate as the distance between the inlet and outlet increases significantly. In general, the empirical equation can get acceptable results.

Besides, as shown in Sect. 7.3, the impact of air inlet location on the first-level floor for buildings below four storeys cannot be ignored. The empirical equation underestimates $V_{\text{in},1}$ as the air inlet moves from the middle of the room to the ceiling. For 1st floor, although the movement of the air inlet to the top shortens the distance between the air inlet and outlet, which affects the buoyancy effect, the air inlet and the window are located at the same level, which undoubtedly reduces the obstruction of airflow through the room. However, the empirical equation only considers the pressure loss across vents and does not take into account the loss during air transport, leading to deviations.

The regression curve predicts the storey correction coefficient based on floor quantities, as shown in Fig. 7.12. The empirical relationship between floors and κ_i can be obtained by regression ($R^2 = 0.9948$),

$$\kappa_i = 1.01\, f^{-0.51} \tag{7.19}$$

where f denotes the total number of floors of the building.

The modified mathematical model is shown in Eq. (7.20). This empirical equation allows the use of a closed-flow-loop method to calculate the floor-by-floor volume flow rate and readily be solved by hand. Although not comparable to the fluid details captured by CFD-based software, the model greatly simplifies the design process and saves numerically computational time. Compared with the experimentally validated numerical simulation model described above, Eq. (7.20) delivers better results in predicting the volume flow rates of a multi-storey solar chimney building.

$$V_i = 1.01 f^{-0.51} \times \left(C_{\text{d}} A_{\text{r},i}\right)^{2/3} \times \left[B_i\left(h_{\text{sc}} - h_{\text{in},i}\right)\right]^{1/3} \tag{7.20}$$

Besides, Eq. (7.20) reflects the relationship between the volume flow rate of each floor and floor quantities. In practice, when the solar radiation intensity and orifice areas of each floor (i.e., air inlet, window, and outlet) are known, the volume flow rate of each floor can be obtained by using the above equation for multi-storey buildings. In the same way, when the known conditions include the required volume flow rate, the inverse calculation can also be used to obtain the corresponding unknowns.

More experiments need to be carried out to apply this model to practical applications. We tried to include as much experimental data as possible from the references for validation. However, very few experimental data of steady flows in buildings ventilated by stack effect are available for a wall solar chimney attached to a multi-storey building with openings. Therefore, no other experimental data can be found in the literature to validate the empirical model.

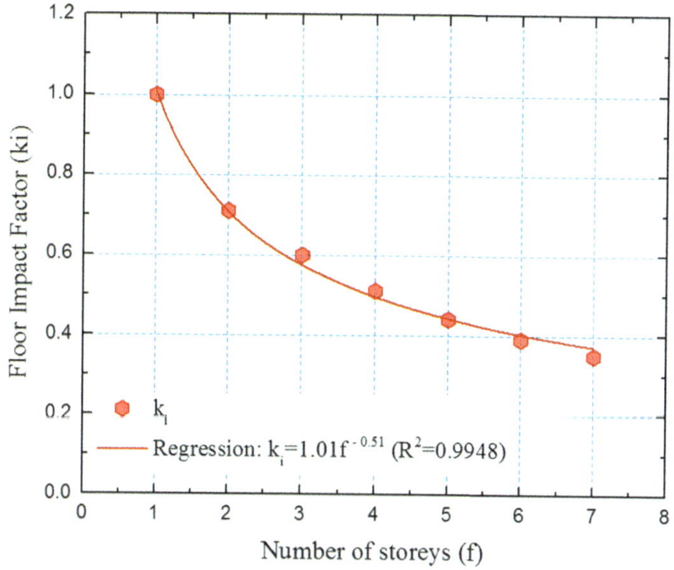

Fig. 7.12 Changing tendency of the average storey correction coefficient (SCC)

7.4 Conclusions

This study theoretically and numerically examined the buoyancy ventilation capacity of a multi-storey building with a solar chimney (SC). A theoretical model was developed to predict the airflow rates for multi-storey SC buildings with identified impacts on storey numbers. The insights gained from this study can assist in providing design strategies for multi-storey buildings. For solar chimneys to increase ventilation capacity, evidence from this study leads to conclusions that:

- The ventilation rate growth trend decreases after exceeding the specific vent size. The air inlet sizes corresponding to the peak discharged volume flow rate ranges from 0.2 to 0.4 m, which is similar to single-zone cases;
- Low-rise buildings' volume flow rates are sensitive to air channel/cavity gaps. As the storey number increases, the gap should increase accordingly to obtain the maximum volume flow rates. Different from the optimal cavity gap (i.e., 0.2–0.3) of a single room reported in the literature, the cavity gaps for peak total flow rates are as follows: 0.5 m for two-storey, 1.0 m for three-storey, 1.2 m for four-storey, and five-storey, 1.5 m for six-storey and seven-storey buildings;
- The larger window area does not significantly improve the volume flow rates, especially the improvement of the upper floors' ventilation is negligible;
- The enhancement of the ventilation effect of lower floors weakens when further increases the floors. In the process of gradually expanding the building from two to seven storeys, the ventilation rate of the top floor shows an exponential declining trend;

- The solar radiation intensity shows a significant positive effect on the ventilation capacity, but it does not contribute as significantly as the cavity gap for multi-storey buildings. The improvement in the maximum volume flow caused by the cavity gap is 45.6% on average, while that caused by solar radiation intensity is only 26.0%;
- Unlike the mathematical model of a single-zone case, the influence of the storey number needs to be considered. Thus, a storey correction coefficient (SCC) expressed by $1.01 f^{-0.51}$ is introduced. This study contributes to a simplified method to obtain a floor-by-floor volume flow rate with a mathematical model (i.e., Eq. (7.20)) for multi-storey SC buildings. Numerical simulations validate the theoretical evidence.

Relying on the experimentally verified FDS numerical simulations, this study obtained an intuitive understanding of the influence of core variables on the ventilation capacity of each floor in medium/low-rise buildings. The closed-loop flow method was used for theoretical derivation to establish a hand-calculated flow equation suitable for medium/low-rise buildings. The simplified mathematical model was compared with the numerical simulation scenarios and obtained good consistency. The derived mathematical model might constitute a practical design and early decision making.

The simplified mathematical model established in this study is suitable for the case of air inlets of identical size on each floor. For simplicity, this study approximates the air within each building zone is at a uniform temperature or well-mixed. The pressure balance used in the ventilation model established in this study only considers the pressure loss at windows, and air inlets and outlets. Therefore, the proposed method's applicability is limited to steady flows in buildings ventilated by the buoyant effect. Further work should shed more light on underlying factors, including dependent variable settings and additional experimental validation.

References

1. Simões N, Manaia M, Simões I (2021) Energy performance of solar and Trombe walls in mediterranean climates. Energy, 121197
2. Noorollahi Y et al (2021) Solar energy for sustainable heating and cooling energy system planning in arid climates. Energy 218:119421
3. Zhang H et al (2021) A critical review of combined natural ventilation techniques in sustainable buildings. Renew Sustain Energy Rev 141:110795
4. Monghasemi N, Vadiee A (2018) A review of solar chimney integrated systems for space heating and cooling application. Renew Sustain Energy Rev 81:2714–2730
5. Zhang TT et al (2016) The application of air layers in building envelopes: a review. Appl Energy 165:707–734
6. Pourshab N et al (2020) Application of double glazed façades with horizontal and vertical louvers to increase natural air flow in office buildings. Energy 200:117486
7. Bachrun AS, Ming TZ, Kurniasih S (2020) Building's solar chimney: ambient obstacle and crosswind in a tropical country. In: EduARCHsia & Senvar 2019 international conference (EduARCHsia 2019). Atlantis Press

8. Miyazaki T, Akisawa A, Kashiwagi TJRE (2006) The effects of solar chimneys on thermal load mitigation of office buildings under the Japanese climate. Renewable Energy 31(7):987–1010
9. Hamdy I, Fikry M (1998) Passive solar ventilation. Renewable Energy 14(1–4):381–386
10. Yang D, Li P (2015) Natural ventilation of lower-level floors assisted by the mechanical ventilation of upper-level floors via a stack. Energy Build 92:296–305
11. Acred A, Hunt GR (2014) A simplified mathematical approach for modelling stack ventilation in multi-compartment buildings. Build Environ 71:121–130
12. Cuce E et al (2019) Sustainable ventilation strategies in buildings: CFD research. Sustain Energy Technol Assess 36:100540
13. Khanal R, Lei C (2011) Solar chimney—a passive strategy for natural ventilation. Energy Build 43(8):1811–1819
14. Khanal R, Lei CW (2015) A numerical investigation of buoyancy induced turbulent air flow in an inclined passive wall solar chimney for natural ventilation. Energy Build 93:217–226
15. Khanal R, Lei CW (2014) An experimental investigation of an inclined passive wall solar chimney for natural ventilation. Sol Energy 107:461–474
16. Khanal R, Lei CW (2014) A scaling investigation of the laminar convective flow in a solar chimney for natural ventilation. Int J Heat Fluid Flow 45:98–108
17. Ren XH et al (2019) Thermal driven natural convective flows inside the solar chimney flush-mounted with discrete heating sources: reversal and cooperative flow dynamics. Renewable Energy 138:354–367
18. Shi L (2018) Theoretical models for wall solar chimney under cooling and heating modes considering room configuration. Energy 165:925–938
19. Zamora B (2021) Determining correlations for solar chimneys in buildings with wind interference: a numerical approach. Sustain Energy Technol Assess 48:101662
20. Shi L (2019) Impacts of wind on solar chimney performance in a building. Energy 185:55–67
21. Jiménez-Xamán C et al (2020) Assessing the thermal performance of a rooftop solar chimney attached to a single room. J Build Eng 31:101380
22. Villar-Ramos M et al (2020) Parametric analysis of the thermal behavior of a single-channel solar chimney. Sol Energy 209:602–617
23. Punyasompun S et al (2009) Investigation on the application of solar chimney for multi-storey buildings. Renewable Energy 34(12):2545–2561
24. Baxevanou C, Fidaros D (2017) Numerical study of solar chimney operation in a two story building. Sustain Synergies Build Urban Scale 38:68–76
25. Gontikaki M et al (2020) Optimization of a solar chimney design to enhance natural ventilation in a multi-storey office building
26. Asadi S et al (2016) The effect of solar chimney layout on ventilation rate in buildings. Energy Build 123:71–78
27. Yang D, Li P (2015) Dimensionless design approach, applicability and energy performance of stack-based hybrid ventilation for multi-story buildings. Energy 93:128–140
28. Chen JL et al (2017) Uncertainty analysis of thermal comfort in a prototypical naturally ventilated office building and its implications compared to deterministic simulation. Energy Build 146:283–294
29. Afonso C, Oliveira A (2000) Solar chimneys: simulation and experiment. Energy Build 32(1):71–79
30. Yang LN, Xu PC, Li YG (2006) Nonlinear dynamic analysis of natural ventilation in a two-zone building: Part A—theoretical analysis. Hvac&R Res 12(2):231–255
31. Wang Y, Wei C (2020) Nonlinear dynamic analysis of solution multiplicity of buoyancy ventilation in two vertically connected open cavities with unequal heights. Complexity
32. Linden PF (1999) The fluid mechanics of natural ventilation. Annu Rev Fluid Mech 31:201–238
33. Shi L et al (2018) Determining the influencing factors on the performance of solar chimney in buildings. Renew Sustain Energy Rev 88:223–238
34. Jiang Y, Chen QY (2001) Study of natural ventilation in buildings by large eddy simulation. J Wind Eng Ind Aerodyn 89(13):1155–1178

35. Shi L, Zhang G (2016) An empirical model to predict the performance of typical solar chimneys considering both room and cavity configurations. Build Environ 103:250–261
36. Lai CM et al (2013) Determinations of the fire smoke layer height in a naturally ventilated room. Fire Saf J 58:1–14
37. Zhang H et al (2021) A wall solar chimney to ventilate multi-zone buildings. Sustain Energy Technol Assess 47:101381
38. McGrattan K et al (2013) Fire dynamics simulator user's guide. NIST Special Publication 1019(6)
39. Jiang Y, Allocca C, Chen Q (2004) Validation of CFD simulations for natural ventilation. Int J Vent 2(4):359–369
40. McGrattan K et al (2013) Fire dynamics simulator technical reference guide volume 1: mathematical model. NIST Special Publication 1018(1):175
41. Obyn S, van Moeseke G (2015) Variability and impact of internal surfaces convective heat transfer coefficients in the thermal evaluation of office buildings. Appl Therm Eng 87:258–272
42. McGrattan KB et al (2000) Fire dynamics simulator—technical reference guide. National Institute of Standards and Technology, Building and Fire Research ….
43. Shi L et al (2019) Interaction effect of room opening and air inlet on solar chimney performance. Appl Therm Eng 159:113877–113888
44. Cheng XD et al (2018) Study on optimizing design of solar chimney for natural ventilation and smoke exhaustion. Energy Build 170:145–156
45. Bouchair A (1989) Solar induced ventilation in the Algerian and similar climates. University of Leeds
46. Cheng X et al (2020) Solar chimney in tunnel considering energy-saving and fire safety. Energy 210:118601
47. Zhou Y et al (2019) Large Eddy Simulation of effectiveness of solid screen on improving natural ventilation performance in urban tunnels. Tunn Undergr Space Technol 86:174–185
48. Arce J et al (2009) Experimental study for natural ventilation on a solar chimney. Renewable Energy 34(12):2928–2934
49. Park D, Battaglia F (2015) Application of a wall-solar chimney for passive ventilation of dwellings. J Solar Energy Eng-Trans ASME 137(6)
50. Zhang TT, Yang HX (2019) Flow and heat transfer characteristics of natural convection in vertical air channels of double-skin solar facades. Appl Energy 242:107–120
51. Bassiouny R, Koura NSA (2008) An analytical and numerical study of solar chimney use for room natural ventilation. Energy Build 40(5):865–873
52. Gan GH (2010) Simulation of buoyancy-driven natural ventilation of buildings—impact of computational domain. Energy Build 42(8):1290–1300
53. Shi L et al (2016) Developing an empirical model for roof solar chimney based on experimental data from various test rigs. Build Environ 110:115–128
54. Andersen KT (1995) Theoretical considerations on natural ventilation by thermal buoyancy. American Society of Heating, Refrigerating and Air-Conditioning Engineers …

Chapter 8
Solar Chimney-Assisted Natural Ventilation in the Tall Buildings

8.1 Introduction

Natural ventilation, esteemed as an energy-conserving approach for building aeration, necessitates reliable methodologies to forecast a building's potential for natural ventilation, playing a pivotal role in governing indoor air parameters. Capitalizing on solar radiation potential, passive solar design strategies could fulfill numerous functions aiming to curtail energy consumption [1, 2]. Adhering to the airflow pattern, the passive ventilation system can markedly realize natural ventilation, space heating/cooling, and air pre-heating [3–6].

A solar chimney, representative of buoyancy-driven ventilation systems, predominantly employs passive solar energy to bolster ventilation. The airflow induced is expedited within an openable air channel and the adjoining occupant spaces [7]. An array of literature has delved into the relationship between solar chimneys' ventilation capacity and design parameters. Comprehensive studies by Khanal and Lei [8–11] primarily targeted the optimization of geometric parameters to enhance the ventilation rates of solar chimneys. Ren et al. [12] investigated the flow dynamics and heat transfer mechanisms of buoyancy within the air channel, unveiling the beneficial impact of discrete heat sources on volume flow rates and heat transfer rates. Shi [13] formulated four theoretical models for natural ventilation in solar chimneys for space cooling/heating, with the empirical results corroborated through experimental validations. In real-world building environments, the influence of external wind is substantial [14]. Shi [15] conducted a numerical and theoretical analysis of the impacts of wind at varying incident angles on the performance of solar chimneys, determining the critical wind speed.

In the domain of airflow network models, Acred and Hunt [16] devised a simplified mathematical technique to model buoyant flow behavior via flow governing equations, formulated by closed-loop pressure balance equations. These were exemplified in a three-storey building with a heated atrium and plenum and numerically solved with equivalent heat inputs on each storey. The model was simplified by

L. Shi and H. Zhang, *Solar Chimney Applications in Buildings*, Green Energy and Technology, https://doi.org/10.1007/978-3-031-45218-5_8

considering solely stack pressure and pressure loss at vents in the pressure term and neglecting air density stratification in each building zone. Zhang et al. executed a CFD simulation with EnergyPlus, comparing the heat transfer and airflow calculation accuracy between the airflow network model and the coupled CFD model. A literature review revealed that the majority of natural ventilation research in residential buildings concentrates on CFD simulation, utilizing indoor airflow distribution or temperatures as evaluation indicators. However, these studies are primarily applicable to certain building types, thereby posing limitations as a reference for designing novel structures [17].

In the context of the thermal model, Punyasompun et al. [18] found that a connected air channel performed better than a separated air channel in diminishing indoor temperature. A key discovery from this study was the numerical model derived from temperature and velocity aspects. This model established a thermal energy balance of the thermal network, aided by an electric analogy for a three-storey solar chimney building. The thermal model offered temperatures for each interior and exterior surface, along with climate data. The wind conditions (speed and direction) from the climate data were concurrently collected and delivered to the airflow simulation, collectively simulating the building's air temperature and airflow rate.

However, the current state of the field exhibits research gaps and limitations concerning the application of solar chimneys in the natural ventilation of high-rise buildings. Despite the growing interest, the potential of solar chimneys in high-rise buildings remains largely unexplored. This dearth of comprehensive research results in a lack of detailed design guidelines and optimization strategies tailored for such applications, thereby inhibiting the broader adoption of solar chimneys in high-rise structures.

Consequently, this study aims to investigate the ventilation efficiency and performance of solar chimneys in high-rise multi-storey buildings, leveraging an experimentally validated numerical model. It explores the commonalities of multi-zone by establishing different scenes. The common airflow problems in the vertically distributed arranged rooms are highlighted.

Building upon this premise, this study delineates three key objectives, each addressing critical aspects of solar chimney application in high-rise buildings.

Firstly, this research aims to identify the critical floor level—the maximum height at which effective natural ventilation can be achieved. This will be instrumental in determining the feasibility and extent of natural ventilation in high-rise buildings, which is fundamental to the incorporation of solar chimneys in such structures.

The second objective is to ascertain the necessary dimensions of the solar chimney to meet the required or average volume flow rate. Such an evaluation will shed light on the optimal size and design of the solar chimney for maximum ventilation efficacy, which is pivotal to the success of the passive cooling strategy in high-rise buildings.

Lastly, the influence of the solar chimney's height and width on the uniform distribution of ventilation rate across different floors will be examined. This assessment aims to ensure a consistent and comfortable indoor environment throughout the building, thereby enhancing the livability and sustainability of high-rise buildings.

By addressing these objectives through Fire Dynamics Simulator (FDS) simulations, this study intends to bridge the identified research gaps and limitations, thereby exploring and validating the potential of solar chimneys in the natural ventilation of high-rise buildings.

8.2 Model Development

8.2.1 Numerical Governing Equations

In this investigation, the Fire Dynamics Simulator (FDS) was utilized, operating on the large eddy simulation (LES) model. This model was employed to resolve energy equations and estimate the volume airflow rates propelled by the defined heat fluxes within multi-storey solar chimney buildings. The FDS is a software product of the U.S. National Institute of Standards and Technology (NIST), known for its reputation and widespread acknowledgment. It numerically addresses a variant of the Navier–Stokes equations suitable for low-speed, thermally driven flow, emphasizing heat and smoke transport. The software's accuracy and effectiveness in predicting natural ventilation have been corroborated by prior research [19–21]. The continuity, momentum, and energy conservation equations utilized in this study have been delineated in our prior research [22].

The continuity equation is provided as

$$(\partial\rho/\partial t + \nabla\cdot\rho\vec{u} = 0) \tag{8.1}$$

The momentum equation is expressed in the following manner:

$$\left(\rho\left((\partial\vec{u})/\,\partial t + (\vec{u}\cdot\nabla)\vec{u}\right) + \nabla p = \rho g + f + \nabla\cdot\vec{\tau}\right) \tag{8.2}$$

The energy equation is noted as

$$(\partial/\partial t(\rho h) + \nabla\cdot\rho h\vec{u} = \mathrm{D}p/\mathrm{D}t - \nabla\cdot q + \varepsilon) \tag{8.3}$$

In these equations, 'ρ' represents the density of air, 't' the time, '$\vec{u}$' the velocity vector, 'p' the pressure, 'f' the external force vector, '$\vec{\tau}$' the viscous stress tensor, 'h' the enthalpy, '$\nabla\cdot q$' the conductive and radiative heat flux, and 'ε' the rate of dissipation wherein kinetic energy is converted into thermal energy through fluid viscosity.

The impact of thermal radiation on the energy conservation equation was modeled using Deardorff's model. An algorithm that uses a default gray gas radiation transfer model is expressed as follows [23]:

$$(s\cdot\nabla I_\lambda(x,s) = \kappa(x,\lambda)[I_b(x) - I(x,s)]) \tag{8.4}$$

In this equation, '$I_b(x)$' represents a Planck function source term, '$I_\lambda(x, s)$' the radiation intensity at wavelength λ, '$\kappa(x, \lambda)$' the absorption coefficient at an 'x' of the wavelength λ, and 's' the unit normal direction vector.

FDS can simulate natural ventilation scenarios using computational grids as an efficient flow-solving technique based on LES for steady and unsteady airflows [24], which filters out small eddies while directly solving conservation equations for larger energy-containing eddies [25].

The building walls are well insulated, with heat conduction between the wall and the air disregarded. The numerical model in this study operates with a constant flow of fresh air supply, ensuring unidirectional flow, implying that the indoor air temperature is infinitely close to the ambient. The convective heat flux in the air channel drives the airflow and produces thermal stratification.

FDS has a default model for describing convective heat transfer between the heated wall and ambient air [25]. The presence of a heated absorption wall makes the cavity temperature consistently higher than that of the rooms. All solid surfaces are considered to have smooth and inert boundary conditions by default to speed up numerical convergence, except for the heated absorption wall. This study specifies the convective heat transfer coefficient as 15.0 W/(m^2 K) [26], and the glazing wall has a transmittance of 0.76 [15]. The values for pressure and ambient temperature are 1 atm and 20 °C.

A series of transient volume flow rates was produced from the numerical simulation. The flow characteristics of the stack were averaged across the vent's cross-section. A duration of 100 s was deemed sufficient to ensure a stable outflow. The airflow volume passing through the air inlets and the outlet can be employed to gage the ventilation capacity of the solar chimney. Consequently, the volume flow rates for the concluding 20 s of the duration were averaged.

The FDS solver uses a dynamically adjusted time step size to solve discrete equations, resulting in computational time savings. The initial default time step size is automatically determined based on the mesh cell size and the flow velocity. Throughout each iteration, time steps across each mesh are synchronized. Additional information regarding these settings can be located in Ref. [27].

8.2.2 *Numerical Scenarios*

To provide a comprehensive understanding of the geometric parameters under consideration, this segment delineates an exemplar setup in the form of a four-storey structure (refer to Fig. 8.1) subjected to solar irradiation. The building envelope features a solar chimney, integrated with a vertical air conduit (d), composed of a heat-absorbing wall coupled with a glazing wall. It is presumed that all solid surfaces within the building are smooth and inert by default. The absorption wall is subject to a steady, uniform heat flux to emulate the absorption of solar radiation.

Individual compartments within the building are interconnected via various flow paths, thereby constructing an integrated airflow network. Each floor communicates

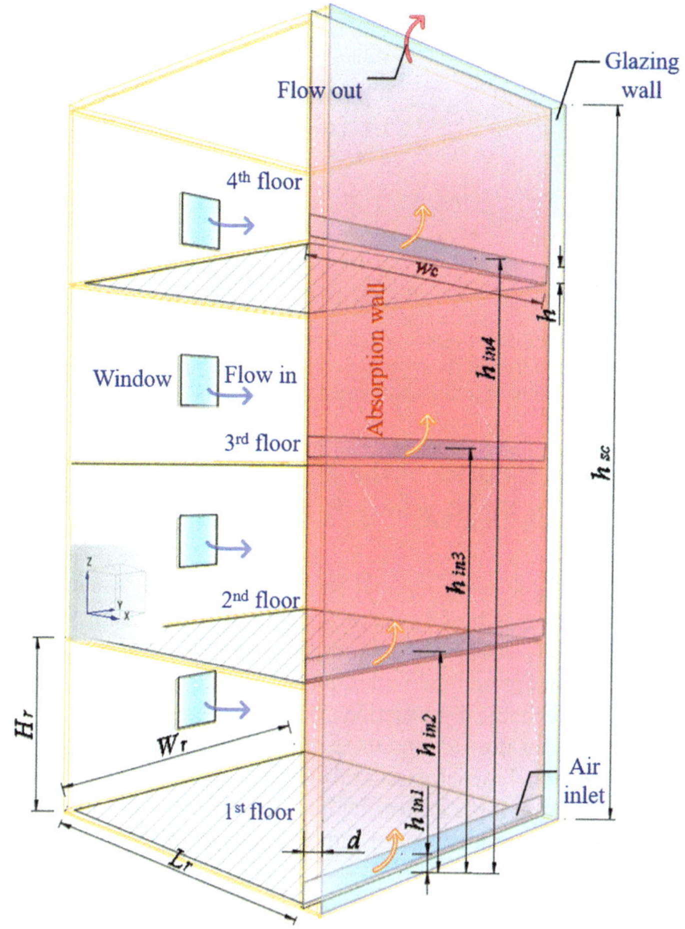

Fig. 8.1 Schematic diagram of an example four-storey building with a solar chimney

with the solar chimney's cavity via a distinct air inlet. All floors share identical internal dimensions, measuring 6.0 m in length (L_r), 6.0 m in width (W_r), and 3.0 m in height (H_r), and are each fitted with a singular air inlet (h) and a square window located centrally on each floor's left wall.

The solar chimney is designed with an air vent at its uppermost section. The vertical distance from the ground level to the center of each air inlet ($h_{in,i}$) signifies the position of the air inlet, which can be adjusted from the floor to the ceiling. The total height of the solar chimney, denoted as hsc, is congruent with the building's height. The width of the chimney, signified as w_c, is equivalent to the building's width.

Table 8.1 shows the variables considered in the numerical study. This study examines tall buildings of four heights.

Table 8.1 Numerical scenarios

Influencing parameters		Values[a]	Interval	Unit
Dimensions	Cavity gap (d)	0.3, 0.6, 1.0, 1.2, 1.5, 1.8	–	m
	Height (h_{sc})	1.5–9.0 m above the building roof	1.5	m
Vents	Window area[b] (A_{open})	1.5 × 1.2	–	m^2
	Air inlet size (h_{in})	0.1–0.5	0.1	m
Floors	Numbers (f)	5, 10, 15, 20	5	–
Environment	Solar intensity (q)	200–1000	200	W/m^2

Notes [a]The variations of the different scenarios are based on the reference case. Only one influencing parameter changes each time, while the other parameters are kept the same as those of the reference case; and [b]window is a square of different dimensions located at the center of the wall

8.3 Flow Characteristics of the Heat Flux Driving Natural Ventilation in Multi-storey Buildings

Solar chimneys function as an integral part of the building's natural ventilation system. Their operation is based on the principle of convection, where warm air rises and is replaced by cooler air, creating a continuous airflow. This mechanism aids in maintaining an optimal indoor air quality and temperature, thereby reducing the dependency on mechanical ventilation systems. This study primarily revolves the evolution of the buoyant flow around the pattern and quantity of airflow throughout rooms into the air channel, which essentially pertains to a multi-floor building model. The scenario presents multiple airflow regimes instigated by the stack effect, which are examined from an intuitive multi-dimensional perspective, specifically for a ten-storey building.

Solar radiation, the key driving factor, infiltrates the vertical air channel through the glazed wall, striking the absorption wall, and subsequently warming the air enclosed within the channel. The ensuing pressure gradient prompts the stale air within rooms to penetrate into the air channel through the inlets, eventually exiting via the outlet of the solar chimney. This work encompasses multiple scenarios, with building heights ranging from five to twenty storeys. However, due to space constraints, the visual simulation results discussed herein predominantly serve as exemplifications of a ten-storey building, unless otherwise specified.

Figure 8.2 graphically elucidates the velocity distributions and flow patterns of a representative ten-storey building under a solar radiation intensity of 600 W/m^2, as viewed from a lateral perspective. The information procured from the Fire Dynamics Simulator (FDS) model pertains to the vertical central plane of the building and the air channel. The air supply rate across each storey experiences variation due to the differential magnitude of the stack effect, which is passively generated by solar buoyancy. As the building height increases, the buoyancy intensity diminishes, thereby reducing the solar chimney's influence on stack pressure and buoyancy-induced airflows. Consequently, the air velocity and the extent of the high-speed zone on the upper floors are considerably lower than those on the lower floors. This results in an observable high-velocity region within the air channel and a distinct disparity in ventilation flow across each floor. The airflow systematically forms a network through the flow path carved by strategically designed vents.

Through the implementation of solar chimneys, high-rise edifices induce high-speed airflow within the confines of each floor in a methodical, step-by-step approach. The air movement initiates from a singular air inlet on each floor and eventually converges within the chimney cavity. Specifically, once the air within the ground floor space has been stabilized and refreshed, a subsequent replacement of the air on the upper floors ensues. This suggests that in high-rise buildings, solar chimneys can create a substantial difference in the internal air pressure, leading to an acceleration in the velocity of the induced airflow. Hence, it can be concluded that the strategic implementation of solar chimneys in high-rise buildings can significantly enhance their ventilation performance, contributing to a healthier and more sustainable built environment.

Contrary to the air velocity distribution within the rooms, the velocity of the air within the cavity demonstrates a consistent increase with height. The air temperature within the air channel surpasses that of the surrounding areas, leading to a significant increase in the air channel's airflow velocity compared to the room. The air adjacent to the absorption wall undergoes heating initially, creating a density difference with the surrounding air, thereby achieving the maximum flow rate at an earlier stage. Furthermore, the cavity airflow is the cumulative effect of the airflow entering the air channel from each floor. Therefore, to quantify the contribution of each floor to the total airflow, the detection of the volume flow rate occurs at the interface of the air inlets and the outlet.

8.4 Influences of Solar Chimney Dimensions on Induced Flow Rate

This section addressed the impacts of the chimney dimensions and solar radiation intensity in solar chimney performance. This study proposed that the optimal geometric parameter value can be perceived as the size that maximizes the ventilation rate induced by buoyancy.

8.4.1 Chimney Height and Gap

The most effective approach for improving the ventilation rate of each floor of a high-rise building and achieving an evenly distributed natural ventilation effect is to simultaneously adjust the height and gap of the air channel. As the number of floors increases, higher chimney heights and corresponding greater gaps are required.

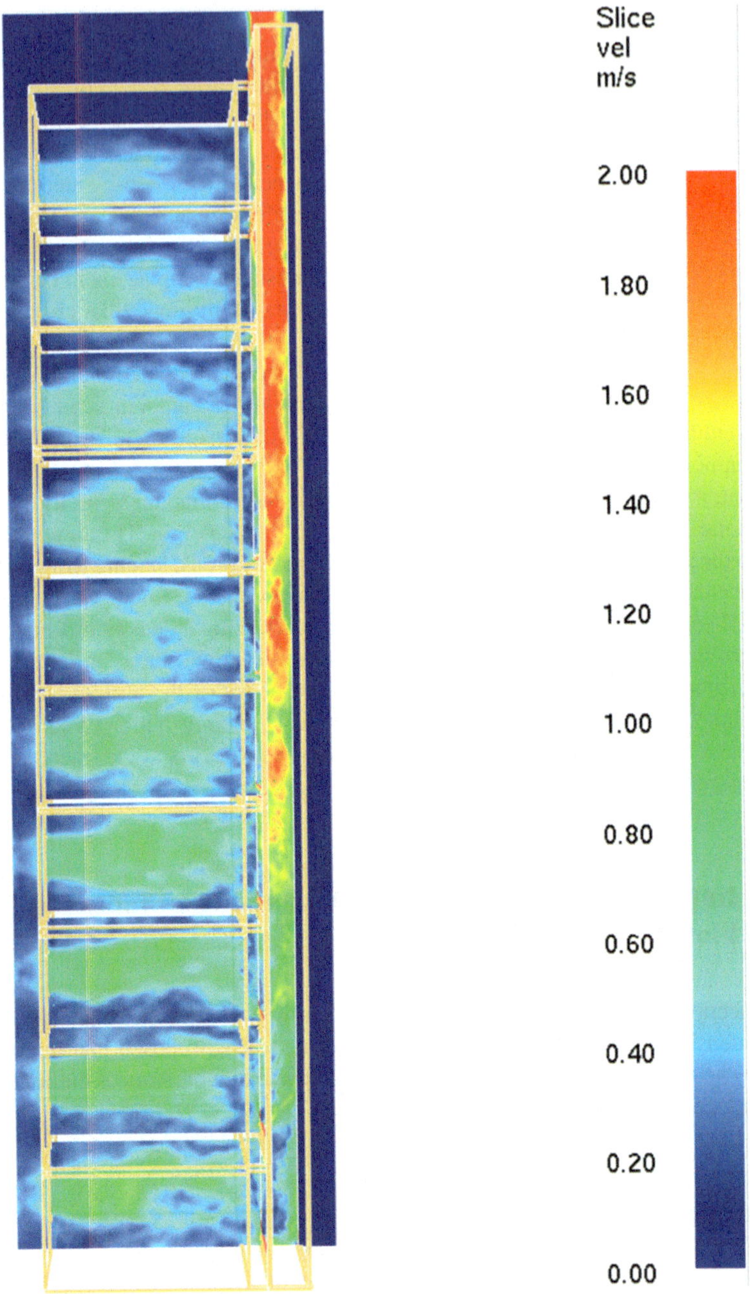

Fig. 8.2 Airflow velocity distribution for the ten-storey case. On the right, a color bar indicates a velocity range of 0–2 m/s

Changing a single parameter can only improve flow on a limited number of floors. As shown in the figure, when the gap of the chimney cavity is maintained at a certain value, simply increasing the height of the chimney cavity can only significantly improve the volume flow of the top floor and the sub-top floor, while the flow of the lower floors hardly changes.

The intricate relationship between the architectural elements of a solar chimney, namely its height and gap, and the subsequent effects on airflow rates within multi-storey buildings serve as key determinants in the performance and efficiency of these structures. An in-depth exploration of this interrelation, as demonstrated in Fig. 8.3, fosters a more profound understanding of the dynamics at play, thereby aiding in the optimal design and utilization of solar chimneys for natural ventilation.

The case studies for a 10-storey and 15-storey building depicted in Fig. 8.3 highlight the significant influence of the gap size on the airflow rates across various floors. An increase in the chimney gap corresponded to a gradual, yet substantial augmentation in the airflow rate. This observation underscores the sensitivity of the airflow enhancement effect to variations in the chimney gap, thereby indicating the importance of careful design consideration in this regard.

On the contrary, the indiscriminate augmentation of the chimney height did not yield an analogous impact on the airflow rates, with the exception of the topmost and

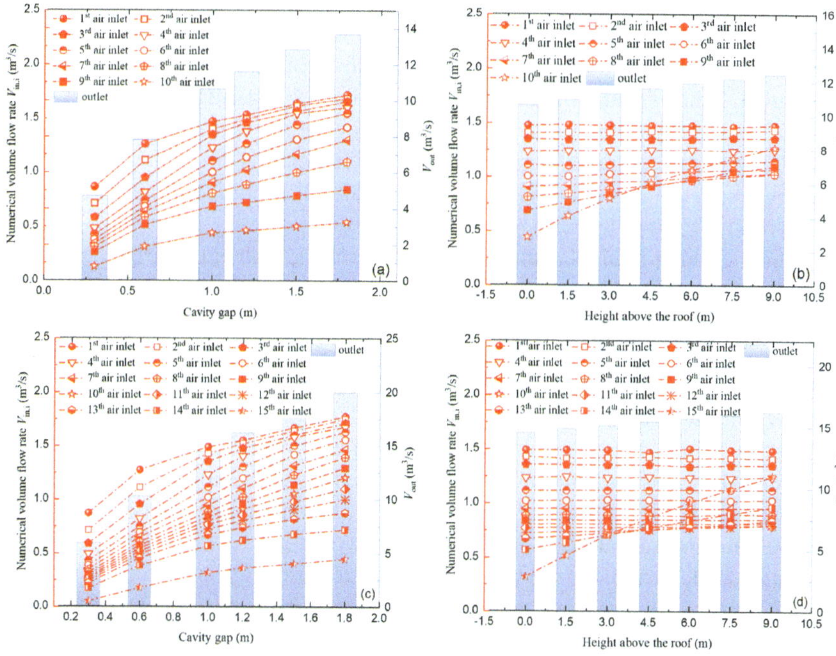

Fig. 8.3 Impact of chimney dimensions on airflow rates: **a** space flow of each floor of the 10-storey building changes with the gap; **b** space flow of each floor of the 10-storey building changes with the chimney height

penultimate floors. This phenomenon can be attributed to the lack of additional air inlets above the top floor space, thereby creating an operational condition reminiscent of the airflow in a single-chambered space influenced by a solar chimney.

As the differential in distance between the outlet of the chimney cavity and the air inlet within the top space augments, it triggers an intensification of the buoyancy effect within the top space. Consequently, this amplified buoyancy effect prompts a marked increase in the flow rate. These observations delineate the intricate balance between the gap and height of the chimney in optimizing the ventilation efficiency.

In order to better understand the impact of chimney dimensions on airflow distribution, this study visualized a case of 10-storey buildings with different chimney heights and gaps, as shown in Fig. 8.6. The analysis elucidates a noteworthy relationship between the height of the chimney and the observed discrepancy in the airflow between the top floor and the remaining floors. While an increase in chimney height catalyzes a more concentrated and accelerated airflow within the air channel, it does not precipitate a substantial escalation in the airflow within the floor spaces. This finding underscores the complex dynamics at play within the solar chimney system and challenges the prevalent assumption that an increase in chimney height invariably translates into enhanced ventilation across all floors.

Conversely, when considering the expansion of the chimney gap, a different pattern emerges. Even though the augmentation of the gap does not necessarily engender high-velocity airflow within the air channel, it does foster a more equitable distribution of airflow. Significantly, this trend toward balanced airflow distribution across each floor space intensifies in tandem with an increase in the gap.

Figure 8.4 shows that an intricate aspect of this investigation involves the exploration of the effects of further increases in chimney height on the airflow rates within the building. As the chimney height increases, it instigates a larger flow rate through the stack. However, this condition could potentially disrupt the formation of natural flows within the lower-level floors.

For tall buildings equipped with numerous floors, the path of the airflow within the cavity expands as the chimney height increases. Consequently, a significant pressure differential becomes necessary to facilitate the expulsion of discharged air out of the stack. This observation suggests that the ventilation strategy, which emphasizes maximizing the flow rate through an increase in chimney height, could be more effective for low-rise buildings. In contrast, its efficiency may decline in the case of high-rise buildings that house an excessively large number of floors.

Nevertheless, this situation may be ameliorated by augmenting the chimney gap. By enhancing the size of the gap, the overall ventilation scheme can be optimized to promote efficient natural flows across all floors.

In sum, while increasing the height of the solar chimney can have certain advantages, it must be approached judiciously, especially in the context of high-rise buildings.

Figure 8.5 presents an intriguing perspective on the optimization of airflow distribution in high-rise buildings. The figure underscores the strategic importance of modulating the gap of the chimney cavity in tandem with the number of floors. This approach suggests that achieving an evenly distributed airflow across each floor may

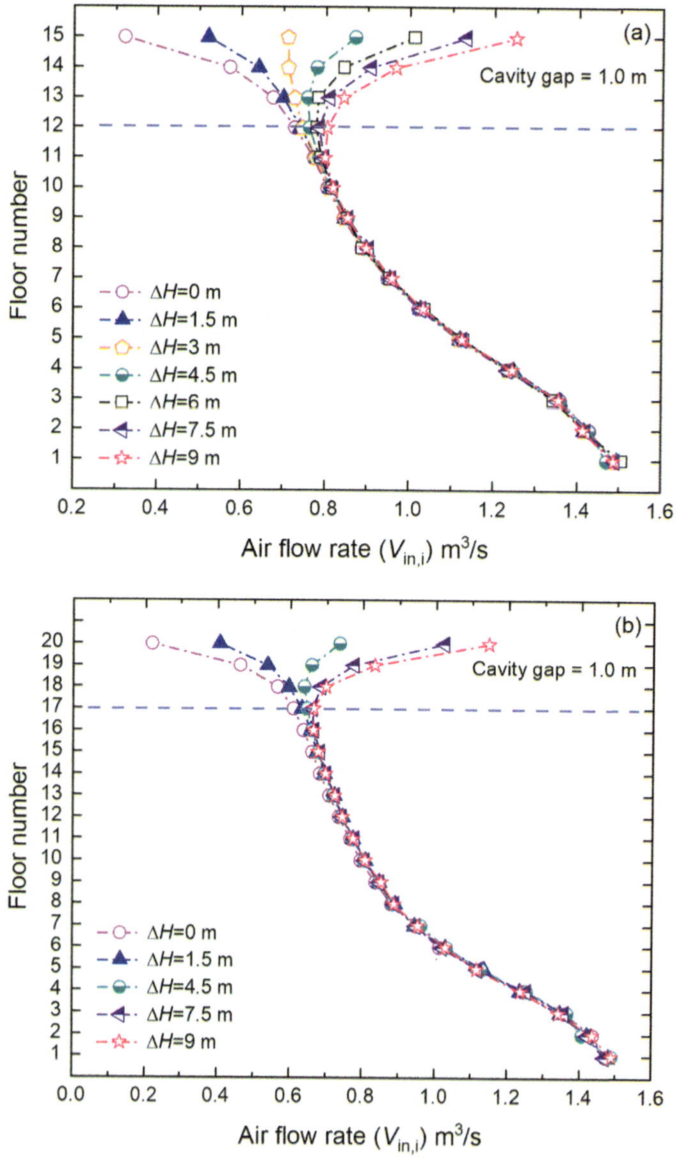

Fig. 8.4 Effect of increasing the chimney height alone on the flow of each floor: **a** fifteen-storey; **b** twenty-storey

necessitate a proportionate increase in the chimney cavity gap with an increase in the number of floors. In other words, the depth of the required cavity gap may be directly proportional to the number of floors.

A closer examination of Fig. 8.5b and d unveils the flow patterns across individual floors. It is noteworthy that altering the cavity gap alone can potentially homogenize the flow rate across each floor, thereby eliminating any significant flow disparities. This study has shown that a larger chimney gap can enhance the induced airflow, which is particularly crucial in multi-storey buildings where ensuring uniform ventilation across floors can be a challenging task.

To summarize, the findings suggest that in the pursuit of an evenly distributed airflow in high-rise buildings, a strategic increase in the gap of the chimney cavity that corresponds to the increase in the number of floors could prove efficacious.

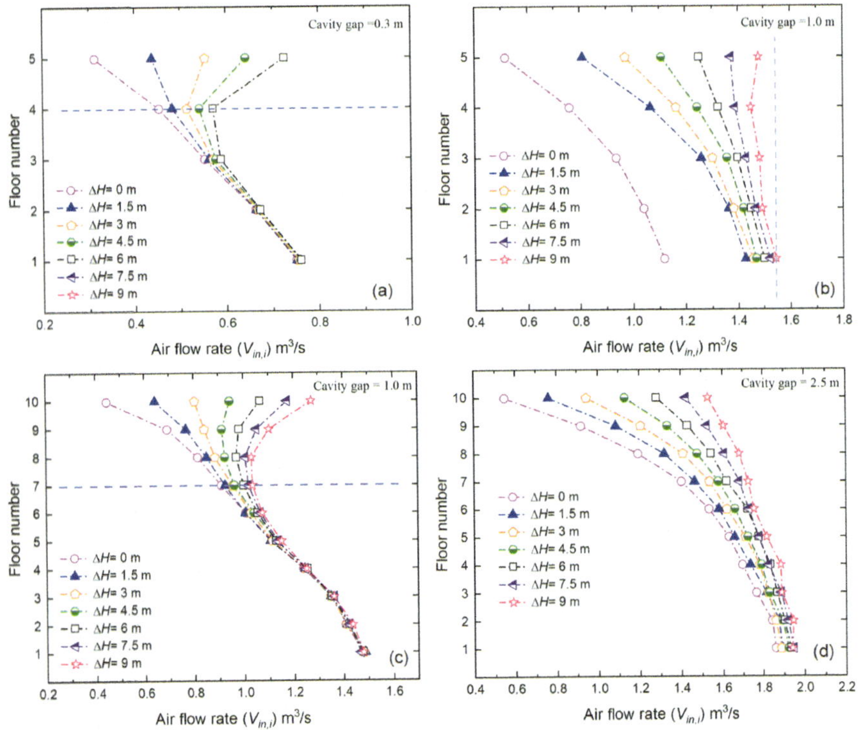

Fig. 8.5 Importance of the cavity gap to facilitate the even distribution of flow on each floor: **a** five-storey with 0.3 m cavity gap; **b** five-storey with 1.0 m cavity gap; **c** ten-storey with 1.0 m cavity gap; and **d** ten-storey with 2.5 m cavity gap

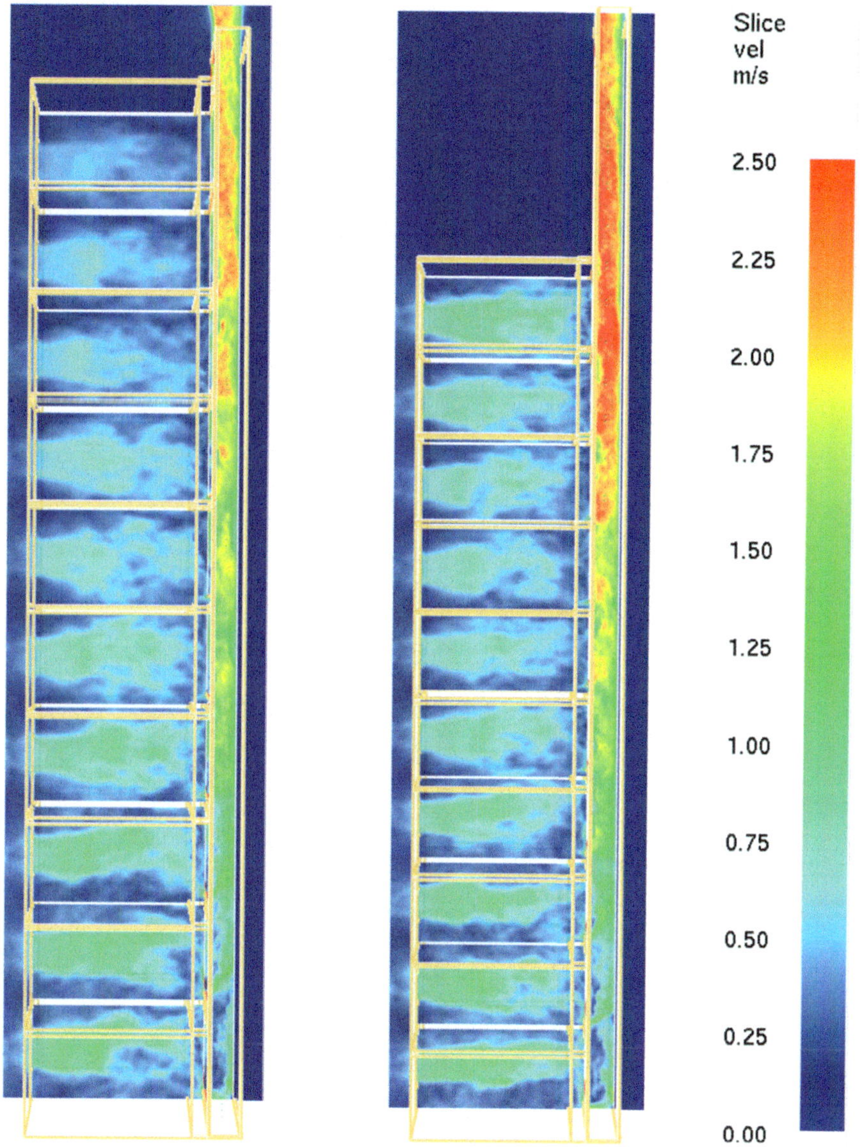

(a) (b)

Fig. 8.6 Influence of solar chimney dimensions on airflow rates in tall buildings: **a** ten-storey with chimney height 1.5 m above the top floor and 1.0 m cavity gap; **b** ten-storey with chimney height 9.0 m above the top floor and 1.0 m cavity gap; **c** ten-storey with 1.2 m cavity gap and chimney height equal to the building height; and **d** ten-storey with 1.8 m cavity gap and chimney height equal to the building height

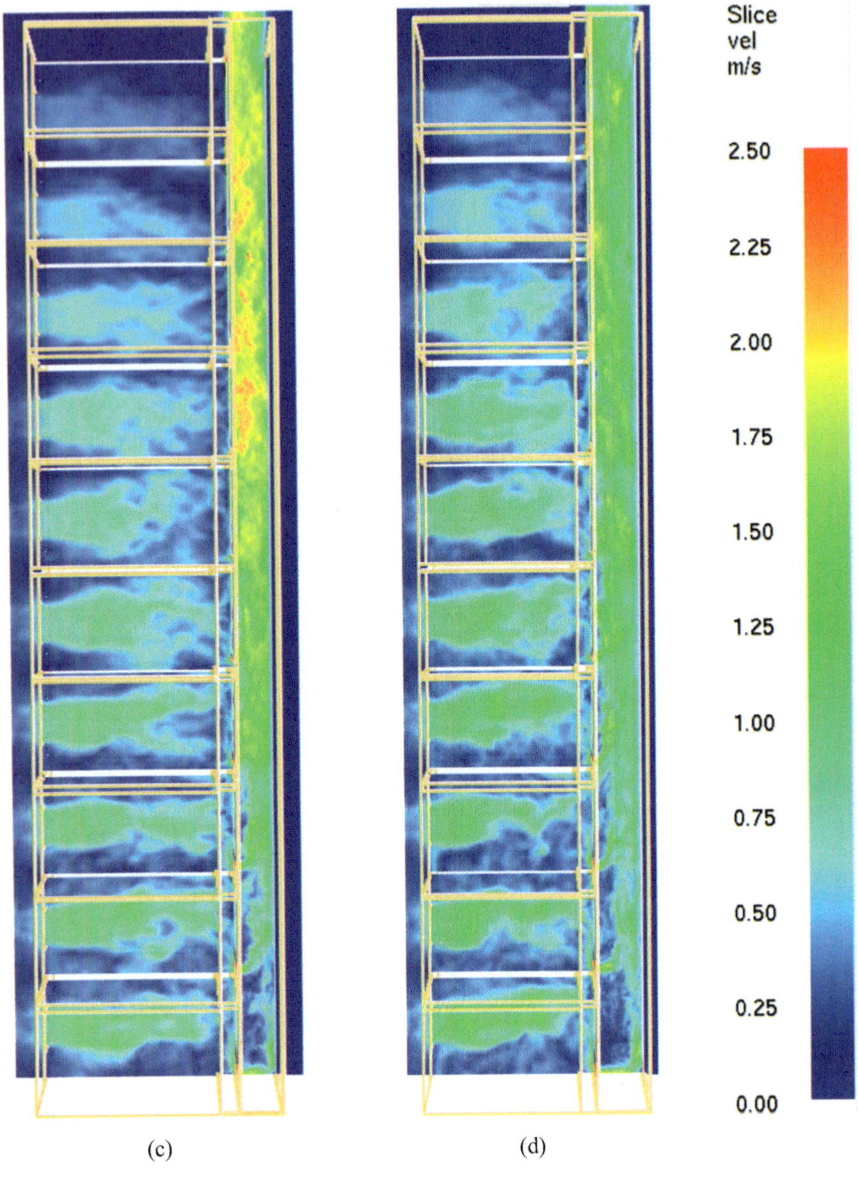

Fig. 8.6 (continued)

8.4.2 Air Inlet Size

Alterations in the inlet area can impact the flow rate captured by the solar chimney. The performance of solar chimneys, when it comes to inducing natural ventilation in

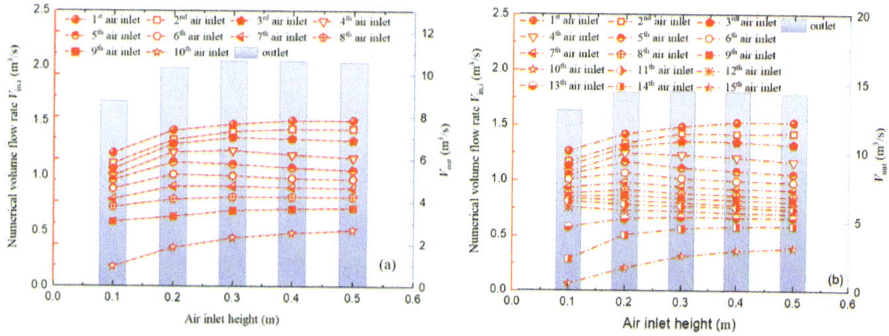

Fig. 8.7 Influences of air inlet height on volume flow rate: **a** ten-storey, **b** fifteen-storey

high-rise buildings, is significantly affected by the size of the air inlet. Past research has underscored the importance of understanding the intricate interaction between the air inlet and room openings such as windows, which ultimately impact solar chimney performance.

These findings are consistent with this work results, where it observed that the flow rate in each floor space reaches a peak and then plateaus when the air inlet height is at 0.3 m or 0.4 m. This behavior was observed in both 10-storey and 15-storey building models, as seen in Fig. 8.7. This correlation underscores the importance of optimizing the air inlet size to enhance the natural ventilation induced by solar chimneys in high-rise buildings.

8.4.3 Input Heat Flux

The efficacy of solar chimneys in inducing natural ventilation in multi-storeyed structures is significantly influenced by the input heat flux. This critical parameter, reflecting the intensity of solar radiation absorbed by the chimney, regulates the stack effect that drives airflow, thereby modulating the ventilation rate.

Similar results were found in the present investigation, which compared the performances of solar chimneys in a 10-storey building and a 15-storey building, shown in Fig. 8.8. Consistent with the findings reported in literature, it was observed that both the flow rate within each floor space and the total induced flux of the solar chimney exhibited an incremental trend with increasing solar heat flux. This underscores the importance of the heat flux as a key determinant in the performance of solar chimneys, particularly in high-rise buildings.

This increasing pattern of airflow rate with heat flux can be comprehended from a thermodynamic perspective. The solar chimney leverages the absorbed solar radiation to heat the air within it, leading to a decrease in air density and a consequent rise in buoyancy. The resultant pressure differential induces an airflow, thereby driving

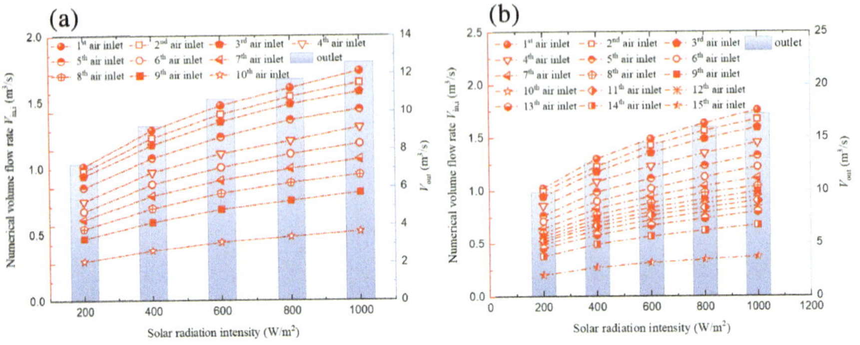

Fig. 8.8 Influence of heat input intensity on airflow rates: **a** ten-storey; **b** fifteen-storey

natural ventilation. Hence, with higher heat flux, more heat is available to warm the air, thereby increasing the airflow rate.

These findings resonate with the theoretical underpinnings of solar chimney operation and corroborate the empirical evidence presented by previous studies. However, it is essential to note that while a higher heat flux could potentially enhance the performance of a solar chimney, other factors such as the design and dimensions of the chimney and the ambient climatic conditions also play a crucial role.

8.5 Conclusions

Based on the research conducted in this study, the performance of solar chimneys in inducing natural ventilation in high-rise buildings is significantly influenced by three main parameters: chimney dimensions, air inlet size, and the input heat flux.

The relationship between the chimney height and gap emerged as a significant determinant in optimizing the ventilation efficiency. Findings showed that an increase in the chimney height can improve the volume flow of the top and penultimate floors, but the flow of the lower floors hardly changes. Conversely, increasing the chimney gap facilitates an even airflow distribution across all floors. This underscores the importance of carefully calibrating both the chimney height and gap to optimize the ventilation rate, especially for high-rise buildings.

The size of the air inlet also plays a crucial role. The study found that the flow rate in each floor space peaks and then plateaus when the air inlet height is at 0.3 m or 0.4 m, emphasizing the need to optimize the air inlet size to enhance the natural ventilation induced by solar chimneys in high-rise buildings.

Lastly, the input heat flux, which reflects the intensity of solar radiation absorbed by the chimney, significantly impacts the stack effect that drives airflow, thereby modulating the ventilation rate. As expected, both the flow rate within each floor space and the total induced flux of the solar chimney exhibited an incremental trend with increasing solar heat flux.

However, it is important to note that while the parameters studied have a profound impact on the performance of solar chimneys, there are other factors such as the design and dimensions of the chimney and the ambient climatic conditions that also play a crucial role. Therefore, a comprehensive approach that takes into account all these factors is needed for the optimal design and utilization of solar chimneys in high-rise buildings.

References

1. Simões N, Manaia M, Simões I (2021) Energy performance of solar and Trombe walls in Mediterranean climates. Energy, 121197
2. Noorollahi Y et al (2021) Solar energy for sustainable heating and cooling energy system planning in arid climates. Energy 218:119421
3. Zhang H et al (2021) A critical review of combined natural ventilation techniques in sustainable buildings. Renew Sustain Energy Rev 141:110795
4. Monghasemi N, Vadiee A (2018) A review of solar chimney integrated systems for space heating and cooling application. Renew Sustain Energy Rev 81:2714–2730
5. Zhang TT et al (2016) The application of air layers in building envelopes: a review. Appl Energy 165:707–734
6. Pourshab N et al (2020) Application of double glazed façades with horizontal and vertical louvers to increase natural air flow in office buildings. Energy 200:117486
7. Cuce E et al (2019) Sustainable ventilation strategies in buildings: CFD research. Sustain Energy Technol Assess 36:100540
8. Khanal R, Lei C (2011) Solar chimney—a passive strategy for natural ventilation. Energy Build 43(8):1811–1819
9. Khanal R, Lei CW (2015) A numerical investigation of buoyancy induced turbulent air flow in an inclined passive wall solar chimney for natural ventilation. Energy Build 93:217–226
10. Khanal R, Lei CW (2014) An experimental investigation of an inclined passive wall solar chimney for natural ventilation. Sol Energy 107:461–474
11. Khanal R, Lei CW (2014) A scaling investigation of the laminar convective flow in a solar chimney for natural ventilation. Int J Heat Fluid Flow 45:98–108
12. Ren XH et al (2019) Thermal driven natural convective flows inside the solar chimney flush-mounted with discrete heating sources: Reversal and cooperative flow dynamics. Renewable Energy 138:354–367
13. Shi L (2018) Theoretical models for wall solar chimney under cooling and heating modes considering room configuration. Energy 165:925–938
14. Zamora B (2021) Determining correlations for solar chimneys in buildings with wind interference: A numerical approach. Sustainable Energy Technol Assess 48:101662
15. Shi L (2019) Impacts of wind on solar chimney performance in a building. Energy 185:55–67
16. Acred A, Hunt GR (2014) A simplified mathematical approach for modelling stack ventilation in multi-compartment buildings. Build Environ 71:121–130
17. Li Y, Li XF (2015) Natural ventilation potential of high-rise residential buildings in northern China using coupling thermal and airflow simulations. Build Simul 8(1):51–64
18. Punyasompun S et al (2009) Investigation on the application of solar chimney for multi-storey buildings. Renewable Energy 34(12):2545–2561
19. Jiang Y, Chen QY (2001) Study of natural ventilation in buildings by large eddy simulation. J Wind Eng Ind Aerodyn 89(13):1155–1178
20. Shi L, Zhang G (2016) An empirical model to predict the performance of typical solar chimneys considering both room and cavity configurations. Build Environ 103:250–261

21. Lai CM et al (2013) Determinations of the fire smoke layer height in a naturally ventilated room. Fire Saf J 58:1–14
22. Zhang H et al (2021) A wall solar chimney to ventilate multi-zone buildings. Sustain Energy Technol Assess 47:101381
23. McGrattan K et al (2013) Fire dynamics simulator user's guide. NIST Special Publication 1019(6)
24. Jiang Y, Allocca C, Chen Q (2004) Validation of CFD simulations for natural ventilation. Int J Vent 2(4):359–369
25. McGrattan K et al (2013) Fire dynamics simulator technical reference guide volume 1: mathematical model. NIST Special Publication 1018(1):175
26. Obyn S, van Moeseke G (2015) Variability and impact of internal surfaces convective heat transfer coefficients in the thermal evaluation of office buildings. Appl Therm Eng 87:258–272
27. McGrattan KB et al (2000) Fire dynamics simulator—technical reference guide. National Institute of Standards and Technology, Building and Fire Research ….

SPRINGER NATURE

GPSR Compliance

The European Union's (EU) General Product Safety Regulation (GPSR) is a set of rules that requires consumer products to be safe and our obligations to ensure this.

If you have any concerns about our products, you can contact us on ProductSafety@springernature.com

In case Publisher is established outside the EU, the EU authorized representative is:

Springer Nature Customer Service Center GmbH
Europaplatz 3
69115 Heidelberg, Germany

The manufacturer's authorised representative in the EU is Springer Nature Customer Service Centre GmbH, Europaplatz 3, 69115 Heidelberg, Germany. If you have any concerns regarding our products, please contact ProductSafety@springernature.com

Printed and bound by CPI Group (UK) Ltd, Croydon, CR0 4YY
01/10/2026
02225919-0018